T0022817

Pseudoscience: A Very Short Introduction

VERY SHORT INTRODUCTIONS are for anyone wanting a stimulating and accessible way into a new subject. They are written by experts, and have been translated into more than 45 different languages.

The series began in 1995, and now covers a wide variety of topics in every discipline. The VSI library currently contains over 700 volumes—a Very Short Introduction to everything from Psychology and Philosophy of Science to American History and Relativity—and continues to grow in every subject area.

Very Short Introductions available now:

ABOLITIONISM Richard S. Newman
THE ABRAHAMIC RELIGIONS Charles L. Cohen
ACCOUNTING Christopher Nobes
ADDICTION Keith Humphreys
ADOLESCENCE Peter K. Smith
THEODOR W. ADORNO Andrew Bowie
ADVERTISING Winston Fletcher
AERIAL WARFARE Frank Ledwidge
AESTHETICS Bence Nanay
AFRICAN AMERICAN HISTORY Jonathan Scott Holloway
AFRICAN AMERICAN RELIGION Eddie S. Glaude Jr
AFRICAN HISTORY John Parker and Richard Rathbone
AFRICAN POLITICS Ian Taylor
AFRICAN RELIGIONS Jacob K. Olupona
AGEING Nancy A. Pachana
AGNOSTICISM Robin Le Poidevin
AGRICULTURE Paul Brassley and Richard Soffe
ALEXANDER THE GREAT Hugh Bowden
ALGEBRA Peter M. Higgins
AMERICAN BUSINESS HISTORY Walter A. Friedman
AMERICAN CULTURAL HISTORY Eric Avila
AMERICAN FOREIGN RELATIONS Andrew Preston
AMERICAN HISTORY Paul S. Boyer
AMERICAN IMMIGRATION David A. Gerber
AMERICAN INTELLECTUAL HISTORY Jennifer Ratner-Rosenhagen
THE AMERICAN JUDICIAL SYSTEM Charles L. Zelden
AMERICAN LEGAL HISTORY G. Edward White
AMERICAN MILITARY HISTORY Joseph T. Glatthaar
AMERICAN NAVAL HISTORY Craig L. Symonds
AMERICAN POETRY David Caplan
AMERICAN POLITICAL HISTORY Donald Critchlow
AMERICAN POLITICAL PARTIES AND ELECTIONS L. Sandy Maisel
AMERICAN POLITICS Richard M. Valelly
THE AMERICAN PRESIDENCY Charles O. Jones
THE AMERICAN REVOLUTION Robert J. Allison
AMERICAN SLAVERY Heather Andrea Williams
THE AMERICAN SOUTH Charles Reagan Wilson
THE AMERICAN WEST Stephen Aron
AMERICAN WOMEN'S HISTORY Susan Ware
AMPHIBIANS T. S. Kemp
ANAESTHESIA Aidan O'Donnell

ANALYTIC PHILOSOPHY Michael Beaney
ANARCHISM Alex Prichard
ANCIENT ASSYRIA Karen Radner
ANCIENT EGYPT Ian Shaw
ANCIENT EGYPTIAN ART AND ARCHITECTURE Christina Riggs
ANCIENT GREECE Paul Cartledge
ANCIENT GREEK AND ROMAN SCIENCE Liba Taub
THE ANCIENT NEAR EAST Amanda H. Podany
ANCIENT PHILOSOPHY Julia Annas
ANCIENT WARFARE Harry Sidebottom
ANGELS David Albert Jones
ANGLICANISM Mark Chapman
THE ANGLO-SAXON AGE John Blair
ANIMAL BEHAVIOUR Tristram D. Wyatt
THE ANIMAL KINGDOM Peter Holland
ANIMAL RIGHTS David DeGrazia
ANSELM Thomas Williams
THE ANTARCTIC Klaus Dodds
ANTHROPOCENE Erle C. Ellis
ANTISEMITISM Steven Beller
ANXIETY Daniel Freeman and Jason Freeman
THE APOCRYPHAL GOSPELS Paul Foster
APPLIED MATHEMATICS Alain Goriely
THOMAS AQUINAS Fergus Kerr
ARBITRATION Thomas Schultz and Thomas Grant
ARCHAEOLOGY Paul Bahn
ARCHITECTURE Andrew Ballantyne
THE ARCTIC Klaus Dodds and Jamie Woodward
HANNAH ARENDT Dana Villa
ARISTOCRACY William Doyle
ARISTOTLE Jonathan Barnes
ART HISTORY Dana Arnold
ART THEORY Cynthia Freeland
ARTIFICIAL INTELLIGENCE Margaret A. Boden
ASIAN AMERICAN HISTORY Madeline Y. Hsu
ASTROBIOLOGY David C. Catling
ASTROPHYSICS James Binney
ATHEISM Julian Baggini
THE ATMOSPHERE Paul I. Palmer
AUGUSTINE Henry Chadwick
JANE AUSTEN Tom Keymer
AUSTRALIA Kenneth Morgan
AUTISM Uta Frith
AUTOBIOGRAPHY Laura Marcus
THE AVANT GARDE David Cottington
THE AZTECS Davíd Carrasco
BABYLONIA Trevor Bryce
BACTERIA Sebastian G. B. Amyes
BANKING John Goddard and John O. S. Wilson
BARTHES Jonathan Culler
THE BEATS David Sterritt
BEAUTY Roger Scruton
LUDWIG VAN BEETHOVEN Mark Evan Bonds
BEHAVIOURAL ECONOMICS Michelle Baddeley
BESTSELLERS John Sutherland
THE BIBLE John Riches
BIBLICAL ARCHAEOLOGY Eric H. Cline
BIG DATA Dawn E. Holmes
BIOCHEMISTRY Mark Lorch
BIOGEOGRAPHY Mark V. Lomolino
BIOGRAPHY Hermione Lee
BIOMETRICS Michael Fairhurst
ELIZABETH BISHOP Jonathan F. S. Post
BLACK HOLES Katherine Blundell
BLASPHEMY Yvonne Sherwood
BLOOD Chris Cooper
THE BLUES Elijah Wald
THE BODY Chris Shilling
THE BOHEMIANS David Weir
NIELS BOHR J. L. Heilbron
THE BOOK OF COMMON PRAYER Brian Cummings
THE BOOK OF MORMON Terryl Givens
BORDERS Alexander C. Diener and Joshua Hagen
THE BRAIN Michael O'Shea
BRANDING Robert Jones
THE BRICS Andrew F. Cooper
BRITISH CINEMA Charles Barr

THE BRITISH CONSTITUTION Martin Loughlin
THE BRITISH EMPIRE Ashley Jackson
BRITISH POLITICS Tony Wright
BUDDHA Michael Carrithers
BUDDHISM Damien Keown
BUDDHIST ETHICS Damien Keown
BYZANTIUM Peter Sarris
CALVINISM Jon Balserak
ALBERT CAMUS Oliver Gloag
CANADA Donald Wright
CANCER Nicholas James
CAPITALISM James Fulcher
CATHOLICISM Gerald O'Collins
CAUSATION Stephen Mumford and Rani Lill Anjum
THE CELL Terence Allen and Graham Cowling
THE CELTS Barry Cunliffe
CHAOS Leonard Smith
GEOFFREY CHAUCER David Wallace
CHEMISTRY Peter Atkins
CHILD PSYCHOLOGY Usha Goswami
CHILDREN'S LITERATURE Kimberley Reynolds
CHINESE LITERATURE Sabina Knight
CHOICE THEORY Michael Allingham
CHRISTIAN ART Beth Williamson
CHRISTIAN ETHICS D. Stephen Long
CHRISTIANITY Linda Woodhead
CIRCADIAN RHYTHMS Russell Foster and Leon Kreitzman
CITIZENSHIP Richard Bellamy
CITY PLANNING Carl Abbott
CIVIL ENGINEERING David Muir Wood
THE CIVIL RIGHTS MOVEMENT Thomas C. Holt
CLASSICAL LITERATURE William Allan
CLASSICAL MYTHOLOGY Helen Morales
CLASSICS Mary Beard and John Henderson
CLAUSEWITZ Michael Howard
CLIMATE Mark Maslin
CLIMATE CHANGE Mark Maslin
CLINICAL PSYCHOLOGY Susan Llewelyn and Katie Aafjes-van Doorn
COGNITIVE BEHAVIOURAL THERAPY Freda McManus
COGNITIVE NEUROSCIENCE Richard Passingham
THE COLD WAR Robert J. McMahon
COLONIAL AMERICA Alan Taylor
COLONIAL LATIN AMERICAN LITERATURE Rolena Adorno
COMBINATORICS Robin Wilson
COMEDY Matthew Bevis
COMMUNISM Leslie Holmes
COMPARATIVE LITERATURE Ben Hutchinson
COMPETITION AND ANTITRUST LAW Ariel Ezrachi
COMPLEXITY John H. Holland
THE COMPUTER Darrel Ince
COMPUTER SCIENCE Subrata Dasgupta
CONCENTRATION CAMPS Dan Stone
CONDENSED MATTER PHYSICS Ross H. McKenzie
CONFUCIANISM Daniel K. Gardner
THE CONQUISTADORS Matthew Restall and Felipe Fernández-Armesto
CONSCIENCE Paul Strohm
CONSCIOUSNESS Susan Blackmore
CONTEMPORARY ART Julian Stallabrass
CONTEMPORARY FICTION Robert Eaglestone
CONTINENTAL PHILOSOPHY Simon Critchley
COPERNICUS Owen Gingerich
CORAL REEFS Charles Sheppard
CORPORATE SOCIAL RESPONSIBILITY Jeremy Moon
CORRUPTION Leslie Holmes
COSMOLOGY Peter Coles
COUNTRY MUSIC Richard Carlin
CREATIVITY Vlad Glăveanu
CRIME FICTION Richard Bradford
CRIMINAL JUSTICE Julian V. Roberts
CRIMINOLOGY Tim Newburn
CRITICAL THEORY Stephen Eric Bronner
THE CRUSADES Christopher Tyerman

CRYPTOGRAPHY Fred Piper and Sean Murphy
CRYSTALLOGRAPHY A. M. Glazer
THE CULTURAL REVOLUTION Richard Curt Kraus
DADA AND SURREALISM David Hopkins
DANTE Peter Hainsworth and David Robey
DARWIN Jonathan Howard
THE DEAD SEA SCROLLS Timothy H. Lim
DECADENCE David Weir
DECOLONIZATION Dane Kennedy
DEMENTIA Kathleen Taylor
DEMOCRACY Bernard Crick
DEMOGRAPHY Sarah Harper
DEPRESSION Jan Scott and Mary Jane Tacchi
DERRIDA Simon Glendinning
DESCARTES Tom Sorell
DESERTS Nick Middleton
DESIGN John Heskett
DEVELOPMENT Ian Goldin
DEVELOPMENTAL BIOLOGY Lewis Wolpert
THE DEVIL Darren Oldridge
DIASPORA Kevin Kenny
CHARLES DICKENS Jenny Hartley
DICTIONARIES Lynda Mugglestone
DINOSAURS David Norman
DIPLOMATIC HISTORY Joseph M. Siracusa
DOCUMENTARY FILM Patricia Aufderheide
DREAMING J. Allan Hobson
DRUGS Les Iversen
DRUIDS Barry Cunliffe
DYNASTY Jeroen Duindam
DYSLEXIA Margaret J. Snowling
EARLY MUSIC Thomas Forrest Kelly
THE EARTH Martin Redfern
EARTH SYSTEM SCIENCE Tim Lenton
ECOLOGY Jaboury Ghazoul
ECONOMICS Partha Dasgupta
EDUCATION Gary Thomas
EGYPTIAN MYTH Geraldine Pinch
EIGHTEENTH-CENTURY BRITAIN Paul Langford
THE ELEMENTS Philip Ball
EMOTION Dylan Evans
EMPIRE Stephen Howe
EMPLOYMENT LAW David Cabrelli
ENERGY SYSTEMS Nick Jenkins
ENGELS Terrell Carver
ENGINEERING David Blockley
THE ENGLISH LANGUAGE Simon Horobin
ENGLISH LITERATURE Jonathan Bate
THE ENLIGHTENMENT John Robertson
ENTREPRENEURSHIP Paul Westhead and Mike Wright
ENVIRONMENTAL ECONOMICS Stephen Smith
ENVIRONMENTAL ETHICS Robin Attfield
ENVIRONMENTAL LAW Elizabeth Fisher
ENVIRONMENTAL POLITICS Andrew Dobson
ENZYMES Paul Engel
EPICUREANISM Catherine Wilson
EPIDEMIOLOGY Rodolfo Saracci
ETHICS Simon Blackburn
ETHNOMUSICOLOGY Timothy Rice
THE ETRUSCANS Christopher Smith
EUGENICS Philippa Levine
THE EUROPEAN UNION Simon Usherwood and John Pinder
EUROPEAN UNION LAW Anthony Arnull
EVANGELICALISM John G. Stackhouse Jr.
EVIL Luke Russell
EVOLUTION Brian and Deborah Charlesworth
EXISTENTIALISM Thomas Flynn
EXPLORATION Stewart A. Weaver
EXTINCTION Paul B. Wignall
THE EYE Michael Land
FAIRY TALE Marina Warner
FAMILY LAW Jonathan Herring
MICHAEL FARADAY Frank A. J. L. James
FASCISM Kevin Passmore
FASHION Rebecca Arnold
FEDERALISM Mark J. Rozell and Clyde Wilcox
FEMINISM Margaret Walters

FILM Michael Wood
FILM MUSIC Kathryn Kalinak
FILM NOIR James Naremore
FIRE Andrew C. Scott
THE FIRST WORLD WAR
Michael Howard
FLUID MECHANICS Eric Lauga
FOLK MUSIC Mark Slobin
FOOD John Krebs
FORENSIC PSYCHOLOGY
David Canter
FORENSIC SCIENCE Jim Fraser
FORESTS Jaboury Ghazoul
FOSSILS Keith Thomson
FOUCAULT Gary Gutting
THE FOUNDING FATHERS
R. B. Bernstein
FRACTALS Kenneth Falconer
FREE SPEECH Nigel Warburton
FREE WILL Thomas Pink
FREEMASONRY Andreas Önnerfors
FRENCH LITERATURE John D. Lyons
FRENCH PHILOSOPHY
Stephen Gaukroger and Knox Peden
THE FRENCH REVOLUTION
William Doyle
FREUD Anthony Storr
FUNDAMENTALISM Malise Ruthven
FUNGI Nicholas P. Money
THE FUTURE Jennifer M. Gidley
GALAXIES John Gribbin
GALILEO Stillman Drake
GAME THEORY Ken Binmore
GANDHI Bhikhu Parekh
GARDEN HISTORY Gordon Campbell
GENES Jonathan Slack
GENIUS Andrew Robinson
GENOMICS John Archibald
GEOGRAPHY John Matthews and
David Herbert
GEOLOGY Jan Zalasiewicz
GEOMETRY Maciej Dunajski
GEOPHYSICS William Lowrie
GEOPOLITICS Klaus Dodds
GERMAN LITERATURE Nicholas Boyle
GERMAN PHILOSOPHY
Andrew Bowie
THE GHETTO Bryan Cheyette
GLACIATION David J. A. Evans
GLOBAL CATASTROPHES Bill McGuire
GLOBAL ECONOMIC HISTORY
Robert C. Allen
GLOBAL ISLAM Nile Green
GLOBALIZATION Manfred B. Steger
GOD John Bowker
GÖDEL'S THEOREM A. W. Moore
GOETHE Ritchie Robertson
THE GOTHIC Nick Groom
GOVERNANCE Mark Bevir
GRAVITY Timothy Clifton
THE GREAT DEPRESSION AND
THE NEW DEAL Eric Rauchway
HABEAS CORPUS Amanda L. Tyler
HABERMAS James Gordon Finlayson
THE HABSBURG EMPIRE
Martyn Rady
HAPPINESS Daniel M. Haybron
THE HARLEM RENAISSANCE
Cheryl A. Wall
THE HEBREW BIBLE AS LITERATURE
Tod Linafelt
HEGEL Peter Singer
HEIDEGGER Michael Inwood
THE HELLENISTIC AGE
Peter Thonemann
HEREDITY John Waller
HERMENEUTICS
Jens Zimmermann
HERODOTUS Jennifer T. Roberts
HIEROGLYPHS Penelope Wilson
HINDUISM Kim Knott
HISTORY John H. Arnold
THE HISTORY OF ASTRONOMY
Michael Hoskin
THE HISTORY OF CHEMISTRY
William H. Brock
THE HISTORY OF CHILDHOOD
James Marten
THE HISTORY OF CINEMA
Geoffrey Nowell-Smith
THE HISTORY OF COMPUTING
Doron Swade
THE HISTORY OF LIFE
Michael Benton
THE HISTORY OF MATHEMATICS
Jacqueline Stedall
THE HISTORY OF MEDICINE
William Bynum
THE HISTORY OF PHYSICS
J. L. Heilbron

THE HISTORY OF POLITICAL THOUGHT Richard Whatmore
THE HISTORY OF TIME Leofranc Holford-Strevens
HIV AND AIDS Alan Whiteside
HOBBES Richard Tuck
HOLLYWOOD Peter Decherney
THE HOLY ROMAN EMPIRE Joachim Whaley
HOME Michael Allen Fox
HOMER Barbara Graziosi
HORMONES Martin Luck
HORROR Darryl Jones
HUMAN ANATOMY Leslie Klenerman
HUMAN EVOLUTION Bernard Wood
HUMAN PHYSIOLOGY Jamie A. Davies
HUMAN RESOURCE MANAGEMENT Adrian Wilkinson
HUMAN RIGHTS Andrew Clapham
HUMANISM Stephen Law
HUME James A. Harris
HUMOUR Noël Carroll
THE ICE AGE Jamie Woodward
IDENTITY Florian Coulmas
IDEOLOGY Michael Freeden
THE IMMUNE SYSTEM Paul Klenerman
INDIAN CINEMA Ashish Rajadhyaksha
INDIAN PHILOSOPHY Sue Hamilton
THE INDUSTRIAL REVOLUTION Robert C. Allen
INFECTIOUS DISEASE Marta L. Wayne and Benjamin M. Bolker
INFINITY Ian Stewart
INFORMATION Luciano Floridi
INNOVATION Mark Dodgson and David Gann
INTELLECTUAL PROPERTY Siva Vaidhyanathan
INTELLIGENCE Ian J. Deary
INTERNATIONAL LAW Vaughan Lowe
INTERNATIONAL MIGRATION Khalid Koser
INTERNATIONAL RELATIONS Christian Reus-Smit
INTERNATIONAL SECURITY Christopher S. Browning
INSECTS Simon Leather
IRAN Ali M. Ansari
ISLAM Malise Ruthven
ISLAMIC HISTORY Adam Silverstein
ISLAMIC LAW Mashood A. Baderin
ISOTOPES Rob Ellam
ITALIAN LITERATURE Peter Hainsworth and David Robey
HENRY JAMES Susan L. Mizruchi
JAPANESE LITERATURE Alan Tansman
JESUS Richard Bauckham
JEWISH HISTORY David N. Myers
JEWISH LITERATURE Ilan Stavans
JOURNALISM Ian Hargreaves
JAMES JOYCE Colin MacCabe
JUDAISM Norman Solomon
JUNG Anthony Stevens
THE JURY Renée Lettow Lerner
KABBALAH Joseph Dan
KAFKA Ritchie Robertson
KANT Roger Scruton
KEYNES Robert Skidelsky
KIERKEGAARD Patrick Gardiner
KNOWLEDGE Jennifer Nagel
THE KORAN Michael Cook
KOREA Michael J. Seth
LAKES Warwick F. Vincent
LANDSCAPE ARCHITECTURE Ian H. Thompson
LANDSCAPES AND GEOMORPHOLOGY Andrew Goudie and Heather Viles
LANGUAGES Stephen R. Anderson
LATE ANTIQUITY Gillian Clark
LAW Raymond Wacks
THE LAWS OF THERMODYNAMICS Peter Atkins
LEADERSHIP Keith Grint
LEARNING Mark Haselgrove
LEIBNIZ Maria Rosa Antognazza
C. S. LEWIS James Como
LIBERALISM Michael Freeden
LIGHT Ian Walmsley
LINCOLN Allen C. Guelzo
LINGUISTICS Peter Matthews
LITERARY THEORY Jonathan Culler
LOCKE John Dunn
LOGIC Graham Priest

LOVE Ronald de Sousa
MARTIN LUTHER Scott H. Hendrix
MACHIAVELLI Quentin Skinner
MADNESS Andrew Scull
MAGIC Owen Davies
MAGNA CARTA Nicholas Vincent
MAGNETISM Stephen Blundell
MALTHUS Donald Winch
MAMMALS T. S. Kemp
MANAGEMENT John Hendry
NELSON MANDELA Elleke Boehmer
MAO Delia Davin
MARINE BIOLOGY Philip V. Mladenov
MARKETING
Kenneth Le Meunier-FitzHugh
THE MARQUIS DE SADE John Phillips
MARTYRDOM Jolyon Mitchell
MARX Peter Singer
MATERIALS Christopher Hall
MATHEMATICAL FINANCE
Mark H. A. Davis
MATHEMATICS Timothy Gowers
MATTER Geoff Cottrell
THE MAYA Matthew Restall and
Amara Solari
THE MEANING OF LIFE
Terry Eagleton
MEASUREMENT David Hand
MEDICAL ETHICS Michael Dunn and
Tony Hope
MEDICAL LAW Charles Foster
MEDIEVAL BRITAIN John Gillingham
and Ralph A. Griffiths
MEDIEVAL LITERATURE
Elaine Treharne
MEDIEVAL PHILOSOPHY
John Marenbon
MEMORY Jonathan K. Foster
METAPHYSICS Stephen Mumford
METHODISM William J. Abraham
THE MEXICAN
REVOLUTION Alan Knight
MICROBIOLOGY Nicholas P. Money
MICROBIOMES Angela E. Douglas
MICROECONOMICS Avinash Dixit
MICROSCOPY Terence Allen
THE MIDDLE AGES Miri Rubin
MILITARY JUSTICE Eugene R. Fidell
MILITARY STRATEGY
Antulio J. Echevarria II
JOHN STUART MILL Gregory Claeys
MINERALS David Vaughan
MIRACLES Yujin Nagasawa
MODERN ARCHITECTURE
Adam Sharr
MODERN ART David Cottington
MODERN BRAZIL Anthony W. Pereira
MODERN CHINA Rana Mitter
MODERN DRAMA
Kirsten E. Shepherd-Barr
MODERN FRANCE
Vanessa R. Schwartz
MODERN INDIA Craig Jeffrey
MODERN IRELAND Senia Pašeta
MODERN ITALY Anna Cento Bull
MODERN JAPAN
Christopher Goto-Jones
MODERN LATIN AMERICAN
LITERATURE
Roberto González Echevarría
MODERN WAR Richard English
MODERNISM Christopher Butler
MOLECULAR BIOLOGY Aysha Divan
and Janice A. Royds
MOLECULES Philip Ball
MONASTICISM Stephen J. Davis
THE MONGOLS Morris Rossabi
MONTAIGNE William M. Hamlin
MOONS David A. Rothery
MORMONISM
Richard Lyman Bushman
MOUNTAINS Martin F. Price
MUHAMMAD Jonathan A. C. Brown
MULTICULTURALISM Ali Rattansi
MULTILINGUALISM John C. Maher
MUSIC Nicholas Cook
MUSIC AND TECHNOLOGY
Mark Katz
MYTH Robert A. Segal
NANOTECHNOLOGY Philip Moriarty
NAPOLEON David A. Bell
THE NAPOLEONIC WARS
Mike Rapport
NATIONALISM Steven Grosby
NATIVE AMERICAN LITERATURE
Sean Teuton
NAVIGATION Jim Bennett
NAZI GERMANY Jane Caplan
NEGOTIATION
Carrie Menkel-Meadow

NEOLIBERALISM Manfred B. Steger and Ravi K. Roy
NETWORKS Guido Caldarelli and Michele Catanzaro
THE NEW TESTAMENT Luke Timothy Johnson
THE NEW TESTAMENT AS LITERATURE Kyle Keefer
NEWTON Robert Iliffe
NIETZSCHE Michael Tanner
NINETEENTH-CENTURY BRITAIN Christopher Harvie and H. C. G. Matthew
THE NORMAN CONQUEST George Garnett
NORTH AMERICAN INDIANS Theda Perdue and Michael D. Green
NORTHERN IRELAND Marc Mulholland
NOTHING Frank Close
NUCLEAR PHYSICS Frank Close
NUCLEAR POWER Maxwell Irvine
NUCLEAR WEAPONS Joseph M. Siracusa
NUMBER THEORY Robin Wilson
NUMBERS Peter M. Higgins
NUTRITION David A. Bender
OBJECTIVITY Stephen Gaukroger
OCEANS Dorrik Stow
THE OLD TESTAMENT Michael D. Coogan
THE ORCHESTRA D. Kern Holoman
ORGANIC CHEMISTRY Graham Patrick
ORGANIZATIONS Mary Jo Hatch
ORGANIZED CRIME Georgios A. Antonopoulos and Georgios Papanicolaou
ORTHODOX CHRISTIANITY A. Edward Siecienski
OVID Llewelyn Morgan
PAGANISM Owen Davies
PAKISTAN Pippa Virdee
THE PALESTINIAN-ISRAELI CONFLICT Martin Bunton
PANDEMICS Christian W. McMillen
PARTICLE PHYSICS Frank Close
PAUL E. P. Sanders
IVAN PAVLOV Daniel P. Todes
PEACE Oliver P. Richmond
PENTECOSTALISM William K. Kay
PERCEPTION Brian Rogers
THE PERIODIC TABLE Eric R. Scerri
PHILOSOPHICAL METHOD Timothy Williamson
PHILOSOPHY Edward Craig
PHILOSOPHY IN THE ISLAMIC WORLD Peter Adamson
PHILOSOPHY OF BIOLOGY Samir Okasha
PHILOSOPHY OF LAW Raymond Wacks
PHILOSOPHY OF MIND Barbara Gail Montero
PHILOSOPHY OF PHYSICS David Wallace
PHILOSOPHY OF SCIENCE Samir Okasha
PHILOSOPHY OF RELIGION Tim Bayne
PHOTOGRAPHY Steve Edwards
PHYSICAL CHEMISTRY Peter Atkins
PHYSICS Sidney Perkowitz
PILGRIMAGE Ian Reader
PLAGUE Paul Slack
PLANETARY SYSTEMS Raymond T. Pierrehumbert
PLANETS David A. Rothery
PLANTS Timothy Walker
PLATE TECTONICS Peter Molnar
PLATO Julia Annas
POETRY Bernard O'Donoghue
POLITICAL PHILOSOPHY David Miller
POLITICS Kenneth Minogue
POLYGAMY Sarah M. S. Pearsall
POPULISM Cas Mudde and Cristóbal Rovira Kaltwasser
POSTCOLONIALISM Robert J. C. Young
POSTMODERNISM Christopher Butler
POSTSTRUCTURALISM Catherine Belsey
POVERTY Philip N. Jefferson
PREHISTORY Chris Gosden
PRESOCRATIC PHILOSOPHY Catherine Osborne
PRIVACY Raymond Wacks
PROBABILITY John Haigh
PROGRESSIVISM Walter Nugent
PROHIBITION W. J. Rorabaugh

PROJECTS Andrew Davies
PROTESTANTISM Mark A. Noll
PSEUDOSCIENCE Michael D. Gordin.
PSYCHIATRY Tom Burns
PSYCHOANALYSIS Daniel Pick
PSYCHOLOGY Gillian Butler and Freda McManus
PSYCHOLOGY OF MUSIC Elizabeth Hellmuth Margulis
PSYCHOPATHY Essi Viding
PSYCHOTHERAPY Tom Burns and Eva Burns-Lundgren
PUBLIC ADMINISTRATION Stella Z. Theodoulou and Ravi K. Roy
PUBLIC HEALTH Virginia Berridge
PURITANISM Francis J. Bremer
THE QUAKERS Pink Dandelion
QUANTUM THEORY John Polkinghorne
RACISM Ali Rattansi
RADIOACTIVITY Claudio Tuniz
RASTAFARI Ennis B. Edmonds
READING Belinda Jack
THE REAGAN REVOLUTION Gil Troy
REALITY Jan Westerhoff
RECONSTRUCTION Allen C. Guelzo
THE REFORMATION Peter Marshall
REFUGEES Gil Loescher
RELATIVITY Russell Stannard
RELIGION Thomas A. Tweed
RELIGION IN AMERICA Timothy Beal
THE RENAISSANCE Jerry Brotton
RENAISSANCE ART Geraldine A. Johnson
RENEWABLE ENERGY Nick Jelley
REPTILES T. S. Kemp
REVOLUTIONS Jack A. Goldstone
RHETORIC Richard Toye
RISK Baruch Fischhoff and John Kadvany
RITUAL Barry Stephenson
RIVERS Nick Middleton
ROBOTICS Alan Winfield
ROCKS Jan Zalasiewicz
ROMAN BRITAIN Peter Salway
THE ROMAN EMPIRE Christopher Kelly
THE ROMAN REPUBLIC David M. Gwynn
ROMANTICISM Michael Ferber
ROUSSEAU Robert Wokler
RUSSELL A. C. Grayling
THE RUSSIAN ECONOMY Richard Connolly
RUSSIAN HISTORY Geoffrey Hosking
RUSSIAN LITERATURE Catriona Kelly
THE RUSSIAN REVOLUTION S. A. Smith
SAINTS Simon Yarrow
SAMURAI Michael Wert
SAVANNAS Peter A. Furley
SCEPTICISM Duncan Pritchard
SCHIZOPHRENIA Chris Frith and Eve Johnstone
SCHOPENHAUER Christopher Janaway
SCIENCE AND RELIGION Thomas Dixon and Adam R. Shapiro
SCIENCE FICTION David Seed
THE SCIENTIFIC REVOLUTION Lawrence M. Principe
SCOTLAND Rab Houston
SECULARISM Andrew Copson
SEXUAL SELECTION Marlene Zuk and Leigh W. Simmons
SEXUALITY Véronique Mottier
WILLIAM SHAKESPEARE Stanley Wells
SHAKESPEARE'S COMEDIES Bart van Es
SHAKESPEARE'S SONNETS AND POEMS Jonathan F. S. Post
SHAKESPEARE'S TRAGEDIES Stanley Wells
GEORGE BERNARD SHAW Christopher Wixson
MARY SHELLEY Charlotte Gordon
THE SHORT STORY Andrew Kahn
SIKHISM Eleanor Nesbitt
SILENT FILM Donna Kornhaber
THE SILK ROAD James A. Millward
SLANG Jonathon Green
SLEEP Steven W. Lockley and Russell G. Foster
SMELL Matthew Cobb
ADAM SMITH Christopher J. Berry
SOCIAL AND CULTURAL ANTHROPOLOGY John Monaghan and Peter Just
SOCIAL PSYCHOLOGY Richard J. Crisp

SOCIAL WORK Sally Holland and Jonathan Scourfield
SOCIALISM Michael Newman
SOCIOLINGUISTICS John Edwards
SOCIOLOGY Steve Bruce
SOCRATES C. C. W. Taylor
SOFT MATTER Tom McLeish
SOUND Mike Goldsmith
SOUTHEAST ASIA James R. Rush
THE SOVIET UNION Stephen Lovell
THE SPANISH CIVIL WAR Helen Graham
SPANISH LITERATURE Jo Labanyi
THE SPARTANS Andrew J. Bayliss
SPINOZA Roger Scruton
SPIRITUALITY Philip Sheldrake
SPORT Mike Cronin
STARS Andrew King
STATISTICS David J. Hand
STEM CELLS Jonathan Slack
STOICISM Brad Inwood
STRUCTURAL ENGINEERING David Blockley
STUART BRITAIN John Morrill
SUBURBS Carl Abbott
THE SUN Philip Judge
SUPERCONDUCTIVITY Stephen Blundell
SUPERSTITION Stuart Vyse
SYMMETRY Ian Stewart
SYNAESTHESIA Julia Simner
SYNTHETIC BIOLOGY Jamie A. Davies
SYSTEMS BIOLOGY Eberhard O. Voit
TAXATION Stephen Smith
TEETH Peter S. Ungar
TELESCOPES Geoff Cottrell
TERRORISM Charles Townshend
THEATRE Marvin Carlson
THEOLOGY David F. Ford
THINKING AND REASONING Jonathan St. B. T. Evans
THOUGHT Tim Bayne
TIBETAN BUDDHISM Matthew T. Kapstein
TIDES David George Bowers and Emyr Martyn Roberts
TIME Jenann Ismael
TOCQUEVILLE Harvey C. Mansfield
LEO TOLSTOY Liza Knapp
TOPOLOGY Richard Earl
TRAGEDY Adrian Poole
TRANSLATION Matthew Reynolds
THE TREATY OF VERSAILLES Michael S. Neiberg
TRIGONOMETRY Glen Van Brummelen
THE TROJAN WAR Eric H. Cline
TRUST Katherine Hawley
THE TUDORS John Guy
TWENTIETH-CENTURY BRITAIN Kenneth O. Morgan
TYPOGRAPHY Paul Luna
THE UNITED NATIONS Jussi M. Hanhimäki
UNIVERSITIES AND COLLEGES David Palfreyman and Paul Temple
THE U.S. CIVIL WAR Louis P. Masur
THE U.S. CONGRESS Donald A. Ritchie
THE U.S. CONSTITUTION David J. Bodenhamer
THE U.S. SUPREME COURT Linda Greenhouse
UTILITARIANISM Katarzyna de Lazari-Radek and Peter Singer
UTOPIANISM Lyman Tower Sargent
VATICAN II Shaun Blanchard and Stephen Bullivant
VETERINARY SCIENCE James Yeates
THE VIKINGS Julian D. Richards
VIOLENCE Philip Dwyer
THE VIRGIN MARY Mary Joan Winn Leith
THE VIRTUES Craig A. Boyd and Kevin Timpe
VIRUSES Dorothy H. Crawford
VOLCANOES Michael J. Branney and Jan Zalasiewicz
VOLTAIRE Nicholas Cronk
WAR AND RELIGION Jolyon Mitchell and Joshua Rey
WAR AND TECHNOLOGY Alex Roland
WATER John Finney
WAVES Mike Goldsmith
WEATHER Storm Dunlop
THE WELFARE STATE David Garland
WITCHCRAFT Malcolm Gaskill
WITTGENSTEIN A. C. Grayling
WORK Stephen Fineman

WORLD MUSIC **Philip Bohlman**
WORLD MYTHOLOGY **David Leeming**
THE WORLD TRADE ORGANIZATION **Amrita Narlikar**
WORLD WAR II **Gerhard L. Weinberg**
WRITING AND SCRIPT **Andrew Robinson**
ZIONISM **Michael Stanislawski**
ÉMILE ZOLA **Brian Nelson**

Available soon:

OBSERVATIONAL ASTRONOMY **Geoff Cottrell**
MATHEMATICAL ANALYSIS **Richard Earl**
HISTORY OF EMOTIONS **Thomas Dixon**

For more information visit our website
www.oup.com/vsi/

Michael D. Gordin

PSEUDOSCIENCE

A Very Short Introduction

OXFORD
UNIVERSITY PRESS

Oxford University Press is a department of the University of Oxford.
It furthers the University's objective of excellence in research, scholarship,
and education by publishing worldwide. Oxford is a registered trade mark of
Oxford University Press in the UK and certain other countries.

Published in the United States of America by Oxford University Press
198 Madison Avenue, New York, NY 10016, United States of America.

© Oxford University Press 2021, 2023
This book was published in hardcover as On the Fringe:
Where Science Meets Pseudoscience (2021)

All rights reserved. No part of this publication may be reproduced,
stored in a retrieval system, or transmitted, in any form or by any means,
without the prior permission in writing of Oxford University Press,
or as expressly permitted by law, by licence, or under terms agreed with
the appropriate reproduction rights organization. Inquiries concerning
reproduction outside the scope of the above should be sent to the
Rights Department, Oxford University Press, at the address above

You must not circulate this work in any other form
and you must impose this same condition on any acquirer

Cataloging-in-Publication data is on file at Library of Congress

ISBN 978-0-19-094442-1

Printed and bound by
CPI Group (UK) Ltd, Croydon, CR0 4YY

Contents

List of illustrations

Preface

Pseudoscience is not a real thing. The term is a negative category, always ascribed to somebody else's beliefs, not to characterize a doctrine one holds dear oneself. People who espouse fringe ideas never think of themselves as "pseudoscientists"; they think they are following the correct scientific doctrine, even if it is not mainstream. In that sense, there is no such thing as pseudoscience, just disagreements about what the right science is. This is a familiar phenomenon. No believer ever thinks she is a "heretic," for example, or an artist that he produces "bad art." Those are attacks lobbed by opponents.

Yet pseudoscience is also real. The term of abuse is deployed quite frequently, sometimes even about ideas that are at the core of the scientific mainstream, and those labels have consequences. If the reputation of "pseudoscience" solidifies around a particular doctrine, then it is very hard for it to shed the bad reputation. The outcome is plenty of scorn and no legitimacy (or funding) to investigate one's theories. In this, "pseudoscience" is a lot like "heresy": if the label sticks, persecution follows.

Sorting out these kinds of debates has traditionally been the domain of philosophy. For religion, we use theology to discriminate between correct and incorrect belief (though that does not mean people agree on the right way to reason theologically). For art, there is aesthetics, and disagreement is rampant there as well. For scientific knowledge, the relevant philosophical domain is *epistemology*, the philosophy of knowledge. Epistemology hits similar roadblocks when it comes to separating science and pseudoscience. This book explores those problems and offers some alternative, nonphilosophical ways to think about the issues. The main approach will be historical: looking at debates from over the past several centuries about what constitutes pseudoscience in order to learn what arguments about the boundaries of acceptable knowledge can tell us about the scientific enterprise as a whole.

This book concerns debates within the natural sciences and not arguments over the humanities and social sciences. Only rarely will medicine come up, and in those instances the focus is on the intersection of medical knowledge with the practices of scientific research. It is hard to exclude phenomena like alternative medicine entirely, but the distinction is nonetheless conceptually significant. The problem of "quackery" in medicine is analogous to "pseudoscience," but sometimes even "false" treatments can make the patient feel better. Efficacy provides a nonepistemological standard in medicine in a way that does not quite happen in science. Tackling pseudoscience separately focuses us on the problem of what counts as *truth*. Some medical claims tackle that head-on, but many others do not.

Understanding how pseudoscience works is an important matter. The problem of reliable knowledge is quite general, ranging from medical treatments to "fake news" to rumors floating among your circle of friends. Thinking about doctrines that have been called "pseudoscience"—creationism, psychical research, UFOlogy, Nazi eugenics, or cold fusion—highlights the dilemmas sharply. What you find in these pages can prove broadly applicable, even if you don't care about Bigfoot.

Chapter 1
The demarcation problem

Any discussion of pseudoscience must start with the so-called demarcation problem. Indeed, without a proposed solution to the demarcation problem—valid or invalid, explicit or implicit—the term *pseudoscience* has no real meaning. If there were a universally recognized and workable demarcation criterion (as one calls a solution to the demarcation problem), then the task of this book would be simple: those doctrines that passed the test would be "science," and those that failed would be "pseudoscience." Alas, the demarcation problem has to date eluded resolution. There are good reasons to think that it will remain a puzzle, which means that debates about what counts as "pseudoscience" will always be with us.

In fact, we have wrestled with the problem of demarcation for as long as domains of knowledge about the natural world have claimed authoritative status. One of the oldest medical writings in the Western tradition, the fifth-century BCE Hippocratic text "On the Sacred Disease," is essentially a demarcation document about how to understand and treat what we now call epilepsy. In the text, the author—conventionally called "Hippocrates," though these documents were likely composed by a variety of authors over a sizable span of time—lambastes "the sort of people we now call witch-doctors, faith-healers, quacks and charlatans." Instead, Hippocrates provides his own theory of the cause of epilepsy, and explains why no faith healer deserves the title of physician. Every claim to scientific authority necessarily implies the exiling of rivals from it.

The basic formulation of the demarcation problem is: how should we distinguish science from pseudoscience? Yet there are really several demarcation problems. There is the core question of epistemology: how do you sift correct knowledge from incorrect claims? Beyond that you also might want to differentiate science from all those domains (art history, theology, gardening) that are "nonscience," or from those things that look an awful lot like science but for some reason do not quite make it. This last set, the imposters, are frequently designated "pseudosciences." Any demarcation criterion worthy of the name ought to be able to distinguish science from them.

The term *demarcation problem* was coined by the philosopher Karl Popper, and his demarcation criterion remains the most commonly invoked among scientists, philosophers, and those undergraduates who have views on this subject. We will start, then, with the philosopher and his criterion of "falsifiability," before elaborating why the criterion fails.

Karl Popper and falsifiability

Karl Popper was born just after the turn of the twentieth century in Vienna, then the capital of the sprawling Austro-Hungarian Empire. By the time he received his doctorate in psychology (not, interestingly, in philosophy) in 1928, he was living in the same city but a very different country: the much smaller republic of Austria. Vienna was home to a vibrant and contentious socialist movement, so he was exposed early on to Marxism but was quickly disillusioned. This was also the birthplace of psychoanalysis, and Popper in the early 1920s volunteered in the clinics of Alfred Adler, who had split with his former mentor, the creator of psychoanalysis, Sigmund Freud. Precocious interest in both theoretical frameworks, and his subsequent rejection of them, were crucial in the later formulation of Popper's philosophy of science.

Philosophy of science was a big deal in Popper's Vienna, and the decade when he was a student saw the flourishing of a group of philosophers called the "Vienna Circle." This group elaborated the dominant philosophy of science of the first half of the twentieth century: *logical empiricism.* Not only did the Vienna Circle and its like-minded peers in Berlin dominate European philosophy of science, but after the rise of National Socialism many of the leading lights (who were either Jewish, or socialist, or both) emigrated to the United States, where they reestablished their school of thought. Popper, though not a member of the Vienna Circle, was likewise thrust into globetrotting, for similar reasons. Although baptized as a Lutheran and a member of a middle-class family, all of his grandparents were Jewish, clouding his future as the annexation of Austria to Hitler's Germany loomed in 1938. Popper emigrated to New Zealand a year before that event and in 1946 moved to London.

Logical empiricism can be usefully understood by examining its component terms. Its advocates are *empiricists* because they believe that sense data constitute our only reliable sources of information about the natural world. Building on centuries of philosophical thought—most notably that of David Hume, the eighteenth-century Scottish philosopher who was especially important for Popper, and Ernst Mach, an Austrian physicist who emphasized the centrality of sense data for the natural sciences—logical empiricists rejected as "metaphysical" any claims about the structure of nature that could not be traced back to sensory observations. Moving beyond Hume and Mach, however, the logical empiricists also stressed the significance of *logical* relations in coherently assembling the shards of reality brought to us through our senses. These logical relations were not necessarily grounded in empirical data themselves, but they were essential to ascertaining nonmetaphysical truths about nature. At first, Popper was quite taken with logical empiricism, but he would diverge with the mainstream of the movement and develop his own framework for understanding scientific thought in *The Logic of Scientific Discovery* (1934, in revised English translation in 1959) and *Conjectures and Refutations* (1963).

Popper claimed to have formulated his initial ideas about demarcation in 1919, when he was seventeen years old. He had "*wished to distinguish between science and pseudo-science*; knowing very well that science often errs, and that pseudo-science may happen to stumble on the truth." That's all very well and good, but how to do it? The results from the British expedition to study the solar eclipse of May 29, 1919, provided the key insight. Astronomers Arthur Eddington and Frank Dyson organized two groups to measure the deflection of starlight around the Sun in order to test a prediction from general relativity, the gravitational theory recently formulated by Albert Einstein. One of Einstein's crucial tests for the theory was that light's path would be bent by strong gravitational fields, such as those surrounding massive bodies like the Sun, and during an eclipse one would be able to measure the precise degree of curvature for light hailing from stars located behind the solar disk. According to Eddington and Dyson, the measured curvature more closely adhered to Einstein's theory than to that predicted by Newtonian gravity. The news made an immediate international sensation, catapulting Einstein to his global celebrity.

Popper was struck by Einstein's prediction for idiosyncratic reasons. "Now the impressive thing about this case," he wrote decades later, "is the risk involved in a prediction of this kind." Had the measurements found Einstein in error, the physicist would have been forced to abandon his theory. Popper built his demarcation criterion around the bravado of wagering against refutation: "One can sum up all this by saying that *the criterion of the scientific status of a theory is its falsifiability, or refutability, or testability.*"

This demarcation criterion is by far the most widely recognized of Popper's philosophical contributions, although it was somewhat of a digression. He first presented it at a lecture sponsored by the British Council at Peterhouse at the University of Cambridge in 1953, and it was later published in *Conjectures and Refutations.* This post–World War II articulation of his demarcation criterion has often obscured the importance of its Austrian origins, though Popper in the lecture stressed its historical roots in post–World War I Vienna.

All demarcation criteria are designed to *exclude* something; although Popper stated that his goal was to explain Einstein's achievement, what he really wanted to do was to show why psychoanalysis and Marxism were not scientific. Those latter theories had been widely understood as "scientific" in his Viennese milieu because of a logical empiricist theory called *verificationism.* According to this view, a theory is scientific if it is verified by empirical data. For Popper, this was grossly insufficient. There was plenty of data that apparently confirmed psychoanalysis, he noted: Freudians could claim that a man with such-and-such characteristics and upbringing would become a homosexual; but they would also claim that someone with the same characteristics who was not homosexual *also* confirmed the theory. In fact, every piece of data about personalities might be another brick in the confirmatory edifice for Freud, just as every event in politics or economics seemingly further confirmed Marxist theories such as the centrality of class conflict in history or the surplus value of labor. To Popper, the logical empiricists were looking at things the wrong way around. The issue was not whether a theory was confirmed—anything might be interpreted as confirming if you formulated the theory flexibly enough. Rather, the point was whether it was possible to *falsify* the theory. Was there any imaginable observation such that, should it be found, Freudians or

Marxists would concede that their theories were false? If the answer was no, these were not sciences. (This is why it is not exactly weighty evidence against Freud and Marx that they fail Popper's criterion; it was literally *designed* to exclude them.) If you claimed to be scientific but could not, as Einstein had, posit conditions under which your theory would be falsified, then you were a pseudoscientist.

The appeal of falsificationism is obvious. It provides a bright line between theories that are scientific and those that can be considered pseudoscientific, and it rewards the boldness that we often like to see exemplified in science. How well does it work?

Falsifying falsificationism

The short answer is: not very. Philosophers of science recognized this almost immediately, for two main reasons. First, it is difficult to determine whether you have actually falsified a theory. This is largely a restatement of one of Popper's objections to verificationism. How do you determine that an observation actually constitutes a confirmation of a theory? Well, you interpret it within its framework, and sometimes those interpretations produce the lamentable distortions that Popper decried. But the same holds true for falsifying a theory. Suppose you did an experiment in your laboratory to test theory *X*, which predicts that under certain conditions your fact-o-meter should register a value of 32.8, and you got a result of 5.63. You have apparently falsified *X*. What do you do? Should you run to the journals and proclaim the death of *X*?

Not so fast. How do you know that your experimental result was accurate? Maybe the reason you did not get the value of 32.8 is that your fact-o-meter malfunctioned, or perhaps you did not perform the experiment under precisely the right conditions. In short, it is rare to have a thumbs-up/thumbs-down result like in the 1919 eclipse expedition. (As a matter of fact, the results of that expedition were more equivocal than Eddington made them seem. It was several years before absolutely incontrovertible results in support of general relativity were obtained, largely by observatories in California.) If any disconfirming result would invalidate the theory that predicted it, then every tenet of modern science would have already been falsified by middle-school science students failing to replicate utterly uncontroversial standard experiments. This is clearly nonsense. While it sounds like a good idea to insist on falsifying observations, it is far from straightforward to determine when precisely this has been done—and that defeats the purpose of having a bright-line standard.

The second problem has to do with the actual demarcations that Popper's criterion gives us. The very minimum we should expect from a demarcation criterion is that it slices the sciences in the right places. We want our criterion to recognize as scientific those theories that are very generally accepted as hallmarks of contemporary science, like quantum physics, natural selection, and plate tectonics. At the same time, we want our criterion to rule out doctrines like astrology and dowsing that are almost universally labeled pseudosciences. Popper's falsifiability standard is not especially helpful in this regard. For starters, it is difficult to present the "historical" natural sciences, such as evolutionary biology, geology, or cosmology—those fields where we cannot "run the tape again" in the laboratory—exclusively in terms of falsifiable claims. Those sciences provide persuasive explanations of nature through the totality of a narrative chain of causal inference rather than a series of empirical yes-no votes. Popper inadvertently excludes important domains of contemporary science.

The situation with inclusion is even worse. The difficulty is sharply expressed by philosopher of science Larry Laudan in an influential article from 1983:

> [Popper's criterion] has the untoward consequence of countenancing as "scientific" every crank claim that makes ascertainably false assertions. Thus flat Earthers, biblical creationists, proponents of laetrile or orgone boxes, Uri Geller devotees, Bermuda Triangulators, circle squarers, Lysenkoists, charioteers of the gods, *perpetuum mobile* builders, Big Foot searchers, Loch Nessians, faith healers, polywater dabblers, Rosicrucians, the-world-is-about-to-enders, primal screamers, water diviners, magicians, and astrologers all turn out to be scientific on Popper's criterion—just so long as they are prepared to indicate some observation, however improbable, which (if it came to pass) would cause them to change their minds.

(Do not worry if many of those doctrines are unfamiliar to you; we will meet most of them in the following pages.) Laudan's critique went further: any bright-line semantic criterion—that is, a formulation that relied on a linguistic test like Popper's—would necessarily fail. He went on to describe the demarcation problem as a "pseudoproblem," a statement that infuriated many philosophers who insisted that it remained a vital question in the philosophy of science. Yet the fact that Laudan was a tad overzealous in his phrasing does not invalidate his point: Popper's criterion does not fringe out many of the doctrines that common usage would demand of it. On the contrary: creationists and UFOlogists often quote Popper to assert that their own positions are scientific and those of their opponents are pseudoscientific.

A more technical examination of Popper reveals that his formulation requires acceding to philosophical positions that are likely uncongenial to his many vocal partisans who readily quote the falsifiability criterion. In his original demarcation article as well as his monumental *Logic of Scientific Discovery*, Popper was explicit that his framework demands that we give up the possibility of ever attaining the truth about nature (or anything else). According to Popper, no scientific theory can, strictly speaking, ever be *true*. The best scientists can achieve is *not yet false*. The existence of atoms, relativity theory, natural selection, the cellular structure of life, gravity, what have you—these are *all* provisional theories awaiting falsification. Popper's is a consistent picture, but it is one that cuts against the intuitions of almost all practicing scientists, philosophers, and the general public.

As comforting as it would be for Popper's clean demarcation criterion to resolve the question of separating science and pseudoscience, both logical analysis and a sociological glance at how scientists and laypeople actually demarcate demonstrate that it does not work. This raises another question: given that the inadequacies of Popper's standard are so evident, why is it so popular?

Popper on trial

The ubiquity of the falsifiability standard is the inadvertent consequence of a legal battle in the United States about "creation science"—a scientized rendering of the Judeo-Christian creation story as depicted in Genesis. A brief examination of this story, which concerns the legality of teaching this doctrine in public schools, introduces some broader themes about the challenges of demarcation and the importance of reflecting on the problem rather than relying on simple (and simplistic) answers.

Controversies over teaching evolution in American public schools simmered during most of the twentieth century, occasionally bursting into open conflagration. The first and most notorious of these is the "Scopes Monkey Trial" of July 1925. Due to the intense boosterism of the town of Dayton, Tennessee, and the immensely successful fictionalization of the story in the stage play (1955, by Jerome Lawrence and Robert E. Lee) and movie (1960, directed by Stanley Kramer and starring Spencer Tracy) *Inherit the Wind*, the story is broadly known. In spring 1925, Tennessee passed the Butler Act, which criminalized the teaching in public schools of human evolutionary descent from primate ancestors. The

American Civil Liberties Union enrolled teacher John Thomas Scopes to knowingly violate the law to test the constitutionality of the ban on Darwinism in court, arguing that by forbidding Darwin's theory because it violated a particular religion's creation story, the Butler Act transgressed the First Amendment of the United States Constitution that prevented the government from establishing a state religion. That Scopes would be convicted was built into the strategy, which centered on appealing the case to the United States Supreme Court. The plan partially worked: Scopes was found in violation of the law and was fined $100.

Scopes appealed to the Tennessee Supreme Court, which set aside the fine on a legal technicality but upheld the constitutionality of the law on the grounds that while it forbade the teaching of evolution, it did not *require* the teaching of any other doctrine of human origins, and thus did not benefit any specific religion. And that is where matters rested. By 1927 thirteen American states had debated similar measures, but only Mississippi and Arkansas enacted them. The Scopes Trial had shown that it was legal to bar the teaching of evolution, but the media hoopla surrounding it had depicted such measures as ridiculous. Although most states did not ban instruction in evolution, they also did not encourage the theory's introduction into the classroom, and there was much regional variation.

Two incidents sparked a reevaluation of the legitimacy of excluding Darwinism from public schools. The first was the Soviet Union's launch of the first artificial satellite, *Sputnik*, on October 4, 1957. The Soviets' success triggered an extensive discussion about whether the United States had fallen behind in science education, and reform proposals were mooted for many different areas, building on the model of the Physical Sciences Study Committee, which had already been impaneled in 1956. The centenary of the publication of Darwin's *On the Origin of Species* (1859), two years after *Sputnik*, prompted biologists to decry that "one hundred years without Darwinism are enough!" The

Biological Sciences Curriculum Study recommended an overhaul of secondary-school education in the life sciences, with Darwinism (and human evolution) given a central place. The ceasefire between the evolutionists and Christian fundamentalists had been broken.

In the 1960s, religious groups countered with a series of laws insisting on "equal time": if Darwinism (or "evolution science") was required subject matter, then it should be balanced with an equivalent theory, "creation science." Those who wanted to challenge the introduction of creationism into school curricula understood that they needed to make their case by arguing about demarcation. If creationism was not science, then it must be religion, and thus could not be taught in public schools, since this would constitute illegitimate state support of religion. (Private schools, then and now, could and do teach what they like.)

Cases from both Arkansas and Louisiana made it to the appellate courts in the early 1980s. The first of these, *McLean v. Arkansas Board of Education*, was a cause célèbre, with a host of expert witnesses sparring over whether Darwinism was science, whether creation science also met the definition of science, and what the limits of the establishment clause of the U.S. Constitution were. A crucial witness for the evolutionists was Michael Ruse, a British philosopher of science then at the University of Guelph in Canada. Ruse testified to several different demarcation criteria and contended that accounts of the origins of humanity based on Genesis could not satisfy them. One of the criteria he floated was Popper's. Judge William Overton, in his final decision in January 1982, cited Ruse's testimony when he invoked that falsifiability was a standard for determining whether a doctrine was science—and that scientific creationism did not meet it. (Ruse walked his testimony back a decade later.) Overton's appellate court decision was expanded by the U.S. Supreme Court in *Edwards v. Aguillard* (1987), the Louisiana case; the result was that Popper's falsifiability was incorporated as a demarcation criterion in a slew of high-school biology texts. No matter that the standard was recognized as bad philosophy; as a matter of legal doctrine it was enshrined. (In his 2005 appellate court decision in *Kitzmiller v. Dover Area School District*, Judge John E. Jones III modified the legal demarcation standards by eschewing Popper and promoting several less sharp but more apposite criteria while deliberating over the teaching of a doctrine known as "intelligent design," a successor of creationism crafted to evade the precedent of *Edwards*.)

Demarcation after Popper

Larry Laudan's 1983 broadside against demarcation as a topic of philosophical inquiry was elicited by his outrage at Ruse's invocation of Popper's falsifiability standard despite its clear flaws. Laudan's rejection of all attempts at demarcation as "pseudoproblems," however, in turn evoked incensed replies from philosophers who noted that demarcation was still a vital topic. More to the point, demarcation is *inevitable*. Scientists have finite time and therefore must select which topics are worth working on and which are not, and this implies some kind of demarcation. Indeed, there seems to be a broad consensus about which doctrines count as fringe, although there remains debate about gray areas. Even conceding that Laudan was correct that bright-line demarcations like Popper's were not tenable, other approaches might prove more successful.

Philosopher (and former professor of biology) Massimo Pigliucci, for example, has suggested that the problem with falsificationism is its one-dimensionality. Although a bright line might not be possible, perhaps we could add more dimensions that corresponded to the heterogeneity of scientific practice. Some sciences, he noted, focused on expanding empirical knowledge; others were more concentrated on deepening our theoretical understanding; some sciences did both, but failure to excel on both axes simultaneously did not disqualify a doctrine from being

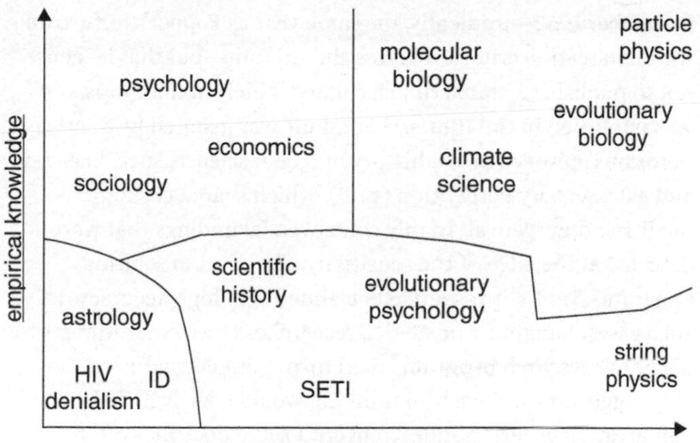

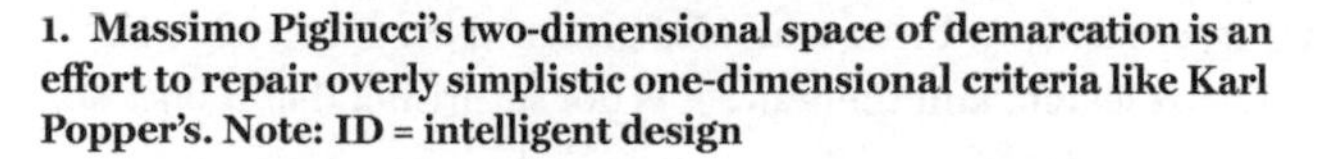

1. Massimo Pigliucci's two-dimensional space of demarcation is an effort to repair overly simplistic one-dimensional criteria like Karl Popper's. Note: ID = intelligent design

"scientific." However, falling too close to the origin of this graph is a reasonably good indication that the subject is not to be considered scientific, and if partisans of one of these doctrines insist on its scientific status, they might find themselves called pseudoscientists. This approach is not flawless, but it avoids some of the pitfalls that beset Popper.

Instead of trying to develop a criterion that will encompass *all* claims to scientific status—an ambition shared by Popper and Pigliucci—you might instead concentrate on what we can think of as "local demarcation criteria": characterizations that encompass groupings of fringe doctrines without claiming to provide a be-all, end-all solution to the demarcation problem.

For example, one influential local demarcation criterion is *pathological science*, a term coined by physical scientist (and 1932 Nobel Laureate in Chemistry) Irving Langmuir in a lecture he gave at General Electric's Knolls Atomic Power Laboratory in

December 1953—ironically, the same year as Popper's lecture on the demarcation criterion across the Atlantic—but that he chose not to publish. (A transcript circulated widely in the 1960s and was published in the 1980s.) Langmuir was inspired by a series of notorious episodes in the history of recent science, such as N-rays and extrasensory perception (ESP), which shared certain qualities: they were all highly controversial findings that were detected at the edge of the sensitivity of current measuring apparatus, yet their researchers claimed very high accuracy. In such cases, Langmuir posited, a researcher's own commitment to his or her research program could turn "pathological": autosuggestion and wishful thinking would take over. This demarcation criterion indeed covers a set of doctrines often labeled "pseudoscience," but it won't do for creationism, or Bigfoot studies, or alchemy. It is a standard designed to specifically rule out ESP research, and therefore it is not surprising that it does so.

In the same fashion, Popper built his falsifiability standard to exclude psychoanalysis, and Ruse and Overton designed theirs to exclude creationism; hence it is no great achievement of their criteria that they successfully do so. All demarcation criteria have this property: they are built inductively out of specific cases, and therefore cannot hope to cover the whole waterfront of possibilities. For this very reason I cannot offer a blanket demarcation criterion of my own—it would flatten out the diversity of the phenomenon under study.

Instead, we might sort fringe doctrines into "families" that can be usefully analyzed together. Four examples would be: vestigial sciences, which are based on past "legitimate" science that is out of date; hyperpoliticized sciences that are yoked to ideological programs; counterestablishment sciences that replicate the sociological structures of mainstream science; and the lineage of theories that have posited extraordinary powers of mind. These categories often overlap, and you might just as easily label a particular doctrine, such as Mesmerism, as a vestigial science or a counterestablishment science instead of as belonging in the lineage of fringe doctrines of mind. No single taxonomy can classify the entirety of the fringe, because the fringe mirrors the heterogeneity of science itself; hence these four categories—in addition to not being properly sealed off from each other—are far from exhaustive. Reflecting upon the diversity of fringe doctrines can provide tools to understand how mainstream science works, and offer some resources to how to think about the inevitable, and imperfect, task of demarcation.

Chapter 2
Vestigial sciences

Often, people designate a doctrine pseudoscience not so much because of *what* it was as *when* it was. Scientific knowledge is not a fixed repository of information, like a bank of dusty volumes filling a wall of bookcases in your local library. On the contrary, one of science's most notable properties is a tremendous dynamism, and quite a bit of that energy is channeled into refuting or revising past knowledge—this was the feature that inspired Karl Popper's demarcation criterion. No understanding of science is plausible that does not recognize this evolutionary, even revolutionary, quality of what counts as scientific knowledge.

Scientific development means that things that we once believed as correct turn out to be too simplistic, or underdeveloped, or plain false as later researchers gather new evidence and deepen their analyses. This is no less true for present-day research: a lot of what is published as the cutting edge of science this year will turn out to be irrelevant or wrong in the not-too-distant future. Consider two examples. Pluto was defined as a planet when it was discovered in 1930, but on August 24, 2006, the International Astronomical Union reclassified it as a "dwarf planet," thus ending the list of official planets with Neptune. Paleontologists used to think dinosaurs were scaly lizards, but new techniques have revealed that many of these extinct creatures sported feathers, prefiguring their evolution into today's birds. Both innovations shocked those who had grown up with Pluto the planet and naked dinosaurs—this was what they had learned as science in elementary school, so wasn't it reliable knowledge? Yet it is wrong to apprehend science as static. Constant change is not a problem with how science operates; it is science working as normal.

This has an important implication for the category of pseudoscience, especially when you delve into the past. The annals of science are littered with discarded doctrines. In truth, *most* of the history of science consists of its dustbin, and the refuse pile grows daily. If you continue to uphold something after its "sell-by" date—after the consensus of scientists has deemed the proposition no longer productive or correct—you run the risk of being labeled a pseudoscientist. I call such doctrines "vestigial sciences": theories and beliefs that once counted as science but were rejected, so that they have morphed today into being classed as pseudosciences.

Among such vestigial sciences are theories almost universally recognized as classic pseudosciences. This venerability holds a double lesson for us. First, such sciences highlight what people in the past held to be the essential characteristics of reliable knowledge, which means we can observe the historical variability of ideas that have been labeled science. Second, historical analysis helps us understand the long process by which these fields were "fringed out" of mainstream science.

Astrology

Astrology is by far the longest-lived doctrine about nature, and it tops many people's lists when asked to name a pseudoscience. Both the longevity and the ubiquity of astrology can obscure rather than reveal. At its most general, we might define astrology as the belief that the positions of the celestial bodies have effects on Earth. Put that way, not only is the belief rather innocuous, it is even true. The Sun's position in the sky corresponds not just with temperature but with the seasons, and—though the mechanism was not well understood until the turn of the eighteenth century—the Moon obviously affects the tides. So much is basic observation. From there, it was but a short leap to include the planets and the fixed stars.

Starting from that general definition, we find that almost every culture has had something that can be dubbed astrology: beyond the Mediterranean region, where the West's astrological tradition consolidated, there are varied astrological systems in East Asia, South Asia, the Middle East, Africa, Latin America, indigenous North America, Scandinavia, and Polynesia. Each of these traditions is in itself heterogeneous, and even in the Western tradition the forms of astrology have evolved significantly over the millennia from their deepest roots in ancient Mesopotamia. It is a long path to the horoscope in your daily newspaper (a practice that began in the 1930s).

Just confining ourselves to the European tradition passed to us from ancient Greece and the medieval Islamic world, one thing is abundantly clear: until roughly the seventeenth century, there was no question that astrology was a science. And not just any science. Astrology was the most empirically grounded and mathematically sophisticated science, its status reflected in the munificent support lavished on it by wealthy patrons. In its elaborate attention to data-gathering, calculation, and prediction—as well as its political importance—it held a position in early modern Europe analogous to economics in the early twenty-first century. Astrology's status was always contested, but it was no less venerable for the fact that people attacked its assumptions and decried its false predictions. (Today, such things are also said about economics.)

One of the best places to appreciate the scientific status of astrology is Italy at the height of the Renaissance. The leading city-states of the Italian peninsula have attracted historians of science in recent decades for many of the same reasons they attracted astrologers in the sixteenth. Trade and banking produced tremendous wealth concentrated in competing city-states, each of which vied for influence and dominance over its peers, buttressed by the tremendous power of the Catholic Church, ensconced in the Holy See in Rome. The nature of the businesses that powered the forceful growth in this region prompted elites to subsidize research into natural knowledge, especially the field then known as "mixed mathematics": an applied domain that today would fall somewhere between physics and engineering. Sailors needed good astronomical observations for navigation, and bookkeepers developed advanced calculational techniques to compute insurance rates. Combined with the humanist study of ancient Greek and medieval Arabic texts, including those related to astrology—such as the *Tetrabiblios*, written in the second century by Claudius Ptolemy (known today for the geocentric astronomical system described in his *Almagest*) and the thirteenth-century Persian scholar Qutb al-Din al-Shirazi—the advanced mathematics and the increasingly accurate observations produced a heady brew.

It is thus no surprise that alongside the glories of painting and sculpture that we associate with the Italian Renaissance, there was a boom in astrology. Every city and every princely court required regular horoscopes to help decide on important events such as weddings or whether to engage the Ottomans in battle. Every scholar of the heavens in this period tried his—they were all men—hand at astrology at one time or another, whether out of pecuniary necessity (as seems to have been the case with Galileo Galilei) or conviction (Johannes Kepler, who served as state astrologer farther north in both Graz and Prague).

What did astrologers do? For the most part, they took the data generated by observational astronomers and used them to produce maps, known as *genitures*, of the heavens at a client's birth. They would later interpret these genitures according to the state-of-the-art knowledge in the discipline. The models

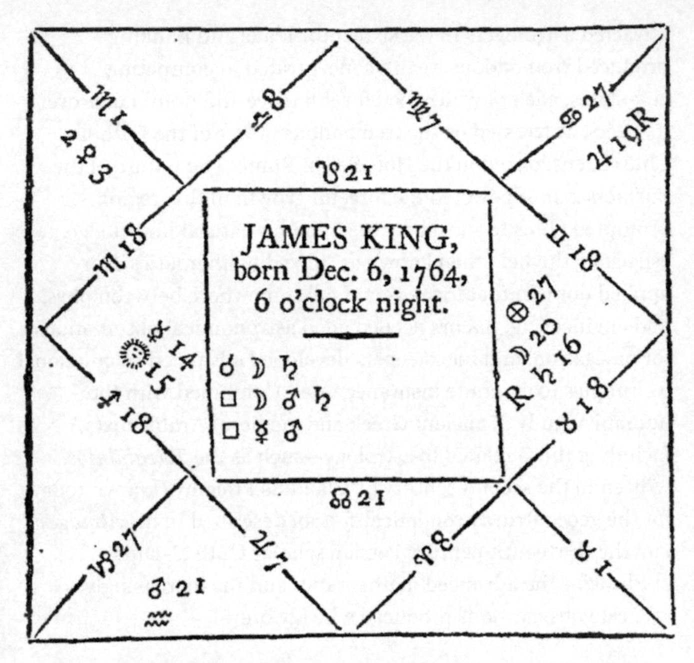

2. This astrological chart ("geniture"), entitled "The Nativity of a dumb Lunatic and Ideot [sic]," was drawn up in 1792.

and interpretations were constantly being refined and older variants were discarded, much as one would expect for any domain of natural philosophy of the day. All these genitures were based on the geocentric cosmology of the sixteenth century.

Earth stands in the middle of the universe, orbited by seven planets: the Moon, the Sun, Mercury, Venus, Mars, Jupiter, and Saturn. (The order of the inner planets was a matter of contention.) These planets moved against the backdrop of the fixed stars, with the Sun completing its circuit in a year; the path it traced against the fixed stars was called the zodiac, divided into twelve segments of 30°, corresponding to the twelve astrological signs. The geniture posited (sometimes implicitly) the signs on the perimeter, with Aries at nine o'clock and then proceeding through the year counterclockwise. The margin was divided into twelve right triangles, representing the "houses" governed by those signs, also beginning at nine o'clock. The houses followed the path of the Sun (the ecliptic) and corresponded to a period of the individual's life; the astrologer placed the symbols for the planets among the houses, determining astronomical positions at the moment of interest—whether it was the birth of an individual or the timing of a particular event—by using star tables, paying particular attention to how the planets stood in relation to the brightest stars of the zodiac. Were they in opposition (180°), trine (120°), quadrature (90°), sextile (60°), or conjunction (0°)? Opposition and quadrature were considered hostile, trine and sextile positive. This was the standard practice of the science for centuries.

Because astrology was ubiquitous, it invited controversy. As long as there has been astrology, people have attacked it as impious, unscientific, or both. Much of the dispute centered around what was known as "judicial astrology": using the stars to predict the course of human events. The distinguished father of the Church Augustine of Hippo disliked the practice, and theological debate raged about the heretical implications: Was God constrained by the heavens? Were genitures a kind of illicit magic? Much less controversial was "natural astrology," which focused on the influence of the heavens on the seasons and on the characters and temperaments of individuals.

Nonetheless, the discipline began to fall out of favor during the seventeenth century, in part related to the development of the heliocentric system—which required significant revisions in the old model (though that did not deter Kepler)—and to religious and cultural transformations. By the late eighteenth century, the science had all but faded, replaced by a positional astronomy grounded in Newtonian mechanics. There was no smoking gun that discredited it, although prominent critics did significant damage; rather, it simply faded from learned discussions across Europe.

It never quite vanished from popular almanacs and cheap print, though. In the mid-nineteenth century, advocates of the new science of meteorology had to defend their predictions against charges of peddling the old astrology in new guise. (The stakes were high for weather forecasters. In 1824, the British Parliament passed a law against vagrancy that criminalized forms of fortune telling, including astrology.) There were also the popular newspapers, which began printing horoscopes for entertainment, although without the precision that a tailored geniture could provide. During the 1970s, a moment when the enthusiasm among scientists for debunking pseudosciences was at high pitch, leading scholars tried to shame the public out of their attachment to the Age of Aquarius, but to no avail. In other parts of the world, such as South Asia, astrology still has a prominent cultural place, used to set auspicious dates for weddings and other important events. Personalized genitures continue to find a reasonable market, but it is not a respectable one in the circles of professional scientists. As a domain of scientific research, it has been resolutely fringed out.

Alchemy

Although alchemy, like astrology, was supported by some princes across Europe, it never fully escaped its association with chicanery and fraud. Nonetheless, the continuity between the doctrines and practices of alchemy and what would later be considered the science of chemistry is striking, quite unlike astrology's relationship to astronomy. Alchemy was decisively fringed away from the mainstream of the study of matter in the early eighteenth century, but this was more a matter of rhetoric than the actual techniques used in the laboratory. Throughout the medieval and early modern periods in Europe, the terms *chemistry* and *alchemy* were used interchangeably, but starting in the eighteenth century we observe a change in France and Britain. Chemists seeking to brand their work as a modern science demonized certain beliefs and practitioners as "alchemical," and the latter fell into ignominy.

We must start with an accurate picture of what "alchemists" did. The Western tradition is not as old as for astrology, dating to third-century CE Greco-Egyptian papyri. Nonetheless, it developed continuously in both medieval Europe and the Islamic world and peaked in the seventeenth century. Since the 1990s, a dedicated group of historians of science have completely revised our understanding of both the doctrines and the practices of medieval and early modern chymistry—a contemporary term they have appropriated in order to encompass both "alchemy" and "chemistry," whose borders were not yet strictly drawn. This blurring was so frequent that many of the titans of natural philosophy such as Isaac Newton and even Robert Boyle ("the father of modern chemistry") expended substantial efforts in alchemical research—work that they performed in secret.

Most people associate alchemy with the quest to turn base metals (lead, iron, etc.) into gold. There were indeed individuals who experimented toward this goal, known as *chrysopoeia*, but it far from exhausted what alchemy's adepts were doing. Much of their work concerned laboratory manipulation of substances, often in the heat of a furnace, in order to understand their transformations, ideally with the consequence of developing some useful substance, such as a medicine. The "philosopher's stone," the red substance that would enable one to turn any base metal into gold, was also sought after as a powerful medicament, and many less omnipotent tinctures and treatments emerged from the search for the stone. At the level of day-to-day operations, most practitioners weighed, assayed, and determined the composition of substances—in short, activities that look a lot like what we now call chemistry.

Even the seemingly most outlandish doctrines—such as the lead-into-gold business—made sense according to the assumptions of the natural philosophy of the time. Shiny metals did not start out that way; they came out of mines as ores that then needed to be purified through heat. When one encountered silver or gold in a mine, the veins were often interpenetrated with other ores. Perhaps the heat of the Earth was cooking the baser ores, slowly making gold out of lesser metals. The alchemist was interested in speeding up this natural process of fermentation or ripening. The obvious economic and medical benefits of manipulating natural substances proved irresistible across the Middle Ages, but it was really in the sixteenth and seventeenth centuries that all forms of chymical research—including chrysopoeia, to be sure, but also study of the properties of metals and the composition of vegetable matter—flourished in Europe.

So how did alchemy develop a reputation for deceit? One explanation is that there *was* a lot of fraud. Traveling con men sought patronage from gullible and cash-strapped lords by promising to fill their treasuries with manufactured gold. They just needed a little bit of gold as a seed to start the process of transforming the lead.... You can see how it worked. This kind of huckster was memorably depicted in the Canon Yeoman's tale in the *Canterbury Tales* of Geoffrey Chaucer, in which the servant tells how his devious master tricked a gullible priest into buying a "recipe" to transform metals after witnessing a series of fraudulent sleights of hand. This reflected a widespread assumption about alchemists, as indicated in severe contemporary injunctions against "conjuring," a crime that targeted witchcraft but included alchemy.

In addition, some of the shady reputation stems from the standard chymical practices of secrecy, indicated by Newton's and Boyle's own habit of keeping their alchemy private. (Boyle had a complementary habit of excoriating it in public.) Secrecy was not called for just because of the proscriptions on "multiplication of metals" or the general bad odor that surrounded the charlatans, but because sincere practitioners understood alchemical knowledge as *powerful*, and therefore requiring protection from unworthy eyes.

This secrecy was enacted through several mechanisms. The first was by training new adepts through apprenticeship. We know, for example, that Boyle learned much of his chymistry (both alchemical and not) through tutoring by George Starkey, a Bermuda-born, Harvard-educated master who was living in London while writing under the pseudonym Eirenaeus Philalethes. Direct master-to-student training meant that chymists could regulate who attained the knowledge and how much of it they received. These tactics derived from artisanal traditions, such as metalsmithing, which shared practical techniques with chymistry. As natural philosophy became more rarefied in this period as an elite activity, these "low" practitioners were excluded, which made the secrecy appear retrograde.

After the invention of print, a second mechanism of restricting the dissemination of alchemical texts was to circulate findings in manuscript, making it easier to control readership. Today, when scientific reputations are built almost entirely on publication, such an attitude seems bizarre, but the mania for publishing all findings as quickly as possible has developed only since the mid-nineteenth century. Avoidance of print was not unusual for philosophers of early modern Europe, who conducted much of their communication through personal correspondence rather than publication. Blaise Pascal and Gottfried Wilhelm von Leibniz, among other luminaries of this period, often circulated their current results in letters to trusted friends, who would in turn copy and forward those writings to the rest of the delimited network. Such practices of secrecy would be cited much later to justify the fringing of alchemists from mainstream science, but in their epoch they were simply common custom.

The third mechanism, associated with printed works, stands apart: the use of *Decknamen* (cover-names) and symbolic language and images. This mode of secrecy dates back to our very earliest alchemical manuscripts. Consider the "second key" of the process of developing the philosopher's stone, from the work attributed to a fifteenth-century alchemist, Basil Valentine. (Valentine actually existed, but he probably did not write the works attributed to him; his name served as a popular pseudonym among sixteenth-century German alchemists.) Each "key" marked a stage in the process, but the procedure was so sensitive that it needed to be protected through encrypted symbols that could only be deciphered by adepts who had received the proper apprenticeship and tutelage. If you knew what you were about, you could read this image. Straightforward is the figure of Mercury in the middle, who likely stands for that element, flanked by a Sun (gold) and Moon (silver). Sometimes, though, an alchemist would *really* mean "philosophical mercury," which they understood to be a purified substance, occasionally interpreted as antimony (Sb on today's periodic table, which of course they didn't have then). Further decipherment is needed for the two other humans, one facing Mercury and one cowering, and the snake and the bird that they wield on their swords. If you do the procedure correctly, translate the cover-names right, you ought to be able to perform the procedure and move on to the next "key."

This might sound like nonsense: perhaps this allegorical picture was just an invention, not intended to represent anything from an actual laboratory. The recent work of historians of alchemy, however, has shown otherwise. For many images, they have been able to decode the symbols so that when they have tried the associated experiment in a modern laboratory with the right chemicals, they have replicated the results the seventeenth-century alchemists claimed. Very few chymical authors boasted that they had personally gotten to gold, but they reported results they had on good authority, and their success in achieving the first four or five stages gave them confidence that the subsequent (and

3. This "second key" from Basil Valentine's 1618 text was meant to encode recipes in its rich, bizarre symbols.

quite challenging) procedures were reliable. As scholars of nature began to adopt different conventions of publication, such presentation styles came to seem preposterous. This was one of the features that made it easier to fringe out the "alchemists" in the eighteenth century.

The very same flamboyance helps account for their revived popularity in later years. By the end of the eighteenth century, alchemy as a tradition seemed to be finished, as the new chemistry of Antoine Lavoisier, based on oxygen and precision measurement, became dominant. Half a century later, however, alchemy enjoyed an unexpected revival. In 1850, Mary Anne Atwood published *A Suggestive Inquiry into the Hermetic Mystery*, which claimed that the coded language of medieval and Renaissance alchemy represented a spiritual search by which the alchemist purified his (or her) own soul. Ethan Allen Hitchcock continued this line of inquiry starting in 1855, and this rendition of spiritual alchemy soon became popular among esoteric religious circles, further distancing the actual early modern practitioners from "science." By the time psychologist Carl Gustav Jung published his highly influential interpretation of alchemy as a quest for spiritual self-fulfillment in 1944, it was a commonplace that the alchemical texts were mystical flights of fancy, not records of laboratory practice. That is why you will be considered a pseudoscientist by the scientific community if you call yourself an alchemist today in any but the most metaphorical sense.

The ubiquity of fringing

Most pseudosciences are vestigial. That is, most doctrines that are so labeled by members of the scientific community in the past and the present do not represent completely new ideas. On the contrary, like astrology and alchemy, they are frequently resurrections or survivals of older ideas that at one time counted as science, but do not any longer. The idea that all species were designed by a divinity was a prominent view among natural historians in the decades and centuries before Charles Darwin's theory of natural selection. If you advocated it in 1820, you would have been doing biology; if you stump for it in 2020, you are a crank. I point this out not in order to rehabilitate these doctrines—though the partial rehabilitation of alchemy by historians is noteworthy—but to make a more general point about how we label and classify fringe theories.

In 1900, the notion that all electromagnetic radiation (including light) traveled on an elastic, weightless medium that penetrated all space, called the *ether*, was not just a common idea, it was virtually an obligatory axiom for practicing physicists. In 1905, Albert Einstein published an article arguing, almost as an aside, that the ether was "superfluous"—you could understand electromagnetism just as well (better, in fact) without it. It took about a decade, but soon the scientific community came to agree: the ether existed no longer. Yet if you survey the advocates of fringe theories today, you will find that "ether physics" forms a prominent subset. From a historical point of view, these people seem to be displaced out of time.

The more you look, the more you see this. *Feng shui*, the ancient Chinese doctrine of geomancy—the study of how to manipulate energy in the surrounding environment—was likewise legitimate knowledge at one point, only to be later discarded. Such fringings are not instant: they unfold over time, sometimes significant stretches of it. By observing these processes, and the regularity of their occurrence, we can understand them as a usual mechanism by which some scientific hypotheses are displaced by their successors. Since there are plenty of discarded doctrines floating under the surface of present knowledge, judging from the future's point of view there are a lot of potential (as well as realized) pseudosciences out there.

Chapter 3
Hyperpoliticized sciences

Consulting a horoscope in the newspaper on occasion does not make one an astrologer, of course, and it does not even require taking the astrologers who draw up those star charts and predictions very seriously. It can simply be an indulgence. You might call it "harmless."

Most scientists do not actively confront doctrines they label "pseudosciences." They can name a few that relate to their own area of expertise, to be sure, but it is a rare individual who takes it upon himself or herself to actively campaign against the offending doctrines and their practitioners. Even those that scientists consider "cranks" are typically "harmless cranks"—not worth wasting time on, and also not doing a great deal of damage.

In the usual course of things, this vision of misguided and mildly irritating benignity is how people think of the category of pseudoscience, which is why they do not think of it that often. But not all such doctrines are harmless. Particularly noteworthy in this regard are a set of positions closely affiliated with repressive political regimes, such as Nazi Germany and Joseph Stalin's reign in the Soviet Union. One might call these doctrines "hyperpoliticized." When the regimes that supported them disintegrated, their advocates largely faded into the woodwork.

The term *hyperpoliticized* requires a little more fleshing out. Some Nazi and Stalinist scientific polices were hugely problematic and destructive, and commentators often attribute the deleterious consequences to their being "political." This is a mischaracterization. All science is at least potentially political, whether on the smaller-scale dimension of hierarchies and prestige within a discipline, or on the broader stage as the recipient of government funds and mobilized to project an (often benign or even salutary) vision of a country as being "pro-science." During the Cold War, the United States State Department and the Central Intelligence Agency sponsored numerous science projects as part of a concerted policy to demonstrate that American democracy was more objective and suited to the production of knowledge than the Soviet Union. When Thabo Mbeki's government in South Africa criticized the import of antiretroviral drugs intended to curb the spread of HIV and to mitigate South Africa's AIDS crisis, this was a political move; so was international pressure against Mbeki's denial of a linkage between HIV and AIDS. That the science concerning antiretrovirals represents the overwhelming consensus of epidemiologists and virologists does not make their promotion less political.

The problem with the Nazi and Soviet cases is not that the science was "political" or even "politicized"—climate science and knowledge of reproductive health are often politicized today—but that they were *hyperpoliticized* purely as arms of a particular political ideology. Although examples from dictatorships are most prominent, such hyperpoliticization can also take place in democratic polities, and many salient characteristics are shared across highly divergent political regimes. Since these doctrines are politically salient, politicians (of varied ideological persuasions) frequently invoke them as negative examples in order to castigate their opponents. They are also, precisely because of the immense political energy required to maintain them, quite rare.

Aryan Physics

"In reality, as with everything that man creates, science is determined by race or by blood," wrote German physicist Philipp Lenard in his four-volume textbook *Deutsche Physik*—which can be translated as "German Physics," but I will render it "Aryan Physics"—in August 1935. Lenard believed that there were different "physicses" for different peoples, although not all of them were equally scientific. Take, for example, Albert Einstein's theories of special and general relativity or the quantum theory of the atom that had been transforming Lenard's own science in recent decades. These were not proper Aryan Physics, but wrong-headed "Jewish" physics. "Jewish 'physics' is therefore only an illusion and a degenerate manifestation of fundamental Aryan physics. [. . .] The unspoiled German national spirit seeks depth; it seeks *theoretical foundations consistent with nature*, and irrefutable knowledge of the cosmos." Jewish physics was the opposite of all of these things, and Lenard wanted it extirpated from German education in Adolf Hitler's Third Reich.

The content of Aryan Physics was elucidated by Lenard and his close collaborator Johannes Stark not only in this textbook but in a number of articles in the popular press as well as in targeted academic journals created for this purpose. The central core of Aryan Physics was to designate other physical theories as pseudoscientific. Lenard and Stark valorized painstaking experimentation and sharp intuition of physical phenomena, which made them suspicious of the highly mathematized theoretical physics that dominated the day. That some of the chief advocates of those latter theories were Jewish—principally Einstein, who was a lightning rod for Nazi attacks due to his pacifism, socialism, and Zionism—provided further anti-Semitic motivation for their campaign. (They ignored the fact that most of the scientists who worked in these theoretical domains were neither Jewish nor foreign.) What Lenard and Stark were left with was basically Newtonian physics and classical Maxwellian electromagnetism, both identified with Aryan (albeit British) progenitors. Although its primary content was negative, a denigration of leading physical theories, there was an affirmative, nostalgic aspect to Aryan Physics, resembling somewhat the patterns we have seen with vestigial sciences. (Of course, Newton's and James Clerk Maxwell's physics were not rendered obsolete by the new developments, only constrained in their domains of applicability.)

There is an irony in Lenard and Stark leading the charge against quantum theory, as they were both associated with its early development. Both were distinguished experimentalists who had performed precision measurements demonstrating the breakdown of classical theories and thereby helped ground the new science of the atom. Lenard was born in what was then known as Pressburg in the Kingdom of Hungary (today it is Bratislava, the capital of Slovakia), and he established his reputation through detailed investigation of cathode rays, the radiation that emanated from the negative electrodes of partially evacuated glass tubes. His innovations enabled him to study the absorption of cathode rays—identified in 1897 with a new, negatively charged fundamental particle, the electron—and especially their similarity to what was ejected from metals with the absorption of ultraviolet light (the "photoelectric effect"). Lenard's measurements earned him the Nobel Prize in Physics in 1905, the same year a patent clerk in the Swiss capital of Bern named Albert Einstein offered a theoretical quantum explanation of the phenomenon. (Einstein received his own Nobel for this work in 1921.) Stark's achievements were analogous. In 1913, he measured the spectra emitted by atoms when subject to strong electric fields, noting a splitting of the resulting lines, which was soon dubbed the "Stark effect." He won the Nobel in Physics in 1919 for this work, incorporated as a central empirical foundation for quantum models of the atom.

Their scientific reputations as titans of German experimental physics made it hard to completely ignore Lenard and Stark as they began airing their anti-Semitic views about the decay of physics in the early 1920s. It became harder still after Adolf Hitler's seizure of the German government in 1933. The two physicists were what was known as "old fighters" (*alte Kämpfer*) in Hitler's movement, having supported the National Socialist Party before it came to power; indeed, they had even backed Hitler's failed 1923 Beer Hall Putsch in Munich. With the rise of the Third Reich and a purging and reformatting of all German institutions in line with Nazi ideas, the two contrarian physicists thought they would be able to rebuild German physics in their image.

Thanks to their political ineptness, a poor match with their ideological fervor, it was not to be. Starting in 1936, Lenard and Stark pushed aggressively for the extirpation of relativity and quantum theory from German pedagogy, but encountered two roadblocks. The first was the stonewalling of Bernhard Rust, the Reich's education minister, who did not care for outsiders meddling in his bailiwick. Sharper was the opposition by other leading German physicists, such as Werner Heisenberg, who insisted that quantum physics and relativity were simply *physics* and could not be junked without jeopardizing work physicists were doing on behalf of the Nazi military. Lenard and Stark attempted to counter Heisenberg by labeling him a "white Jew" (i.e., an Aryan race traitor) in the leading periodical of the SS, but after interrogation Heisenberg was saved from a worse fate by the intervention of Heinrich Himmler, whose mother played bridge with Heisenberg's. Aryan Physics was never really established in the Third Reich.

This is not to say that Nazi policies did not have significant effects on German science. First of all, the Civil Service Law of April 1933 entailed the dismissal of all Jews or socialists who were in civil service positions, including university professorships. This sparked a mass dispersion of some of the leading lights of German science and had a decisive impact on both the Germany they fled and the Allied countries (principally the United States) they emigrated to. The demographic character of the German scientific workforce was completely altered. On the other hand, the German state poured enormous sums into war-related scientific work, ranging from rockets to the pollination patterns of bees (to improve agricultural production). These buoyed up science probably more than hyperpoliticized doctrines like Aryan Physics damaged it—though not enough, most likely, to compensate for the refugees.

Finally, and most significantly, German medicine, physiology, anthropology, genetics, and other human sciences were mobilized in pursuit of Nazi racial policies up to and including the Final Solution—the genocide of European Jews. The Roma and Sinti, the mentally and physically disabled, homosexuals, and other categories were also swept up in a public health net that at first sterilized and then murdered millions before the war was over. This, too, was hyperpoliticized science, ended only when the regime fell in May 1945.

Lysenkoism

While these events took place in Hitler's Germany, a different form of hyperpoliticized science was unfolding in the discipline of genetics in Joseph Stalin's Soviet Union. The central figure in this story was a Ukrainian-born agronomist named Trofim Denisovich Lysenko, who—building upon the horticultural theories and practices of the elderly plant breeder Ivan Michurin—claimed he could change the hereditary properties of plants by exposing them to environmental stressors. In this sense, his doctrine was a development in the neo-Lamarckian tradition of the inheritance of acquired characteristics, and he called it either "Michurinism," after the venerable (and, after 1935, safely dead) plant breeder, or "agrobiology." Outside of the Communist bloc, especially in the

1950s, his ideas were known as "Lysenkoism," though they were not called this in the Soviet Union until after his death in 1976.

Lysenko owed his spectacular rise from a peasant background to the educational and social opportunities provided by the Bolshevik regime after the Russian Revolution of 1917, and both he and the regime would use this connection to reinforce their separate goals. He first came to the attention of the scientific community with a 1927 article in the official newspaper *Pravda*, when he was just twenty-nine years old, extolling him for proving that one could grow a crop of winter peas in subtropical Azerbaijan. He achieved this feat using a set of procedures that he would come to call "vernalization": subjecting seeds to periods of extreme cold or friction before planting them. This made winter or spring variants of certain crops susceptible to germination in different climates, and in fact had been used as an agronomic technique since the mid-nineteenth century across the Western world. By the early 1930s, Lysenko would claim that vernalization "shattered" the heredity of the plants, thereby enabling these environmental modifications to be passed on to future generations. Classical genetics, a burgeoning field since the rediscovery in 1900 of the midcentury pea-plant experiments of the monk Gregor Mendel, claimed that the units of heredity—called "genes" since 1909—were immutable to environmental modification. The Soviet Union, along with the United States, was one of the most important centers of classical genetics in the 1920s, and an intellectual confrontation was in the offing.

Lysenko continued skillfully using newspapers, which were happy to promote the agronomist's claims in the hopes of motivating peasants during the catastrophic imposition of collectivization in the Soviet countryside (1929–1933). In 1929, Lysenko recruited his peasant father to moisten and chill the seeds of winter wheat and plant them in the spring, and the following year the Ukrainian Commissariat of Agriculture ordered one thousand one-hectare tests. Lysenko claimed success in many of these trials, although the data was limited and statistical analysis essentially absent. He was partially tolerated among the scientists in the Lenin All-Union Academy of the Agricultural Sciences (VASKhNIL), especially by its president, Nikolai Vavilov, a world-renowned geneticist. Lysenko provided some ideological cover in the dangerous Stalinist 1930s, while Michurinists and classical geneticists struggled for primacy. Although by the standards of global science Lysenko's claims were starting to seem old-fashioned, at this point the dispute still bore the characteristics of a disagreement among specialists.

That would change in August 1948, at which point the word *pseudoscience* began to crop up around Lysenko's name in the international press and scientific literature. In line with a series of congresses either planned or actually held in various academic disciplines, VASKhNIL hosted a conference to discuss "The Situation in Biological Science." Its president was now Trofim Lysenko. Nikolai Vavilov had been arrested and sentenced in July 1941 for ostensibly counterrevolutionary activities, another victim of Stalin's state terror campaign. (He died of malnutrition in 1943 in Saratov.) The classical geneticists were on the defensive, and Lysenko lambasted them as "Mendelist-Morganist-Weismannists," after three ideologically problematic titans of the field: Mendel (a Catholic priest), Thomas Hunt Morgan (an American geneticist whose surname happened to match that of the vilified banker, J. P. Morgan), and August Weismann (whose major sin, aside from demonstrating that modifications of body cells do not interact with the hereditary "germ line," was being German and therefore, despite being deceased, implicitly connected with the criminal Nazi uses of genetics).

At this point, Joseph Stalin himself entered the picture. The first recorded mention of Stalin with regard to Lysenko was during a speech the latter gave on February 15, 1935. Upon stumbling in midsentence, Lysenko apologized for being an agronomist, not an orator, and Stalin applauded, saying, "Bravo, Comrade Lysenko, bravo!" The Communist leader kept an eye on the geneticist-Michurinist disagreement, but preferred to keep both groups at loggerheads rather than intervening. Lysenko announced on August 7, 1948, the last day of the VASKhNIL meeting, that this position had changed. While taking questions from the floor, the transcript records Lysenko saying the following: "The question is asked in one of the notes handed to me, What is the attitude of the Central Committee of the Party to my report? *I answer: The Central Committee of the Party has examined my report and approved it.* (Stormy applause. Ovation. All rise.)" Stalin had declared Michurinism/agrobiology as the only legitimate science of heredity. Classical genetics was labeled a pseudoscience in the Soviet Union and its proponents fired, until 1965, when Lysenko finally lost his domination of the field. He had outlasted Stalin by a dozen years and was only felled when prominent academicians (largely from the physical sciences) demanded the Academy of Sciences audit Lysenko's farm in the Lenin Hills, where mismanagement and fraud were rampant. Lysenko died in 1976 in disgrace. In his wake, Soviet biologists worked to catch up in genetics and the new science of molecular biology; the legacy of the criminalization of genetics can still be felt in Russian science today.

Especially in the United States, the Lysenko episode stands as a classic "pseudoscience," a cautionary tale about the evils of Soviet Communism and the dangers of political intervention in science. This interpretation faces a number of empirical difficulties. First, the Soviets, much like the Nazis, heavily invested in science, which as a whole flourished in the Soviet Union. (At the time of the collapse of the Soviet Union in 1991, it boasted the largest national scientific community in the world.) Lysenkoism was atypical. Second, political intervention happens in science all the time, including in democracies: evolution is taught in public schools but teaching creationism there is proscribed; experiments on cloning humans are banned; some sciences are funded and others are not, sometimes on political rather than intellectual

4. In 1948, Trofim Lysenko announced that Joseph Stalin (in the portrait on the right) supported his Michurinist theories of heredity.

grounds. Nevertheless, the Lysenko episode is cited among "pseudosciences" as illustrating the perils of state interference in scientific debates. One wonders what would have happened had Stalin backed the classical geneticists. How would we judge his intervention then?

Eugenics

Hyperpoliticized sciences are often associated with authoritarian regimes. This identification makes sense: such governments often espouse strong ideological positions that extend to natural science, and they deploy coercive mechanisms to enforce orthodoxy. Though such episodes in the history of pseudoscience seem to occur more frequently under these kinds of governments, democratic or liberal regimes are not immune. The case of eugenics in the United States demonstrates this clearly enough.

"Eugenics," stemming from Greek roots meaning "well born," was coined by Francis Galton, a cousin of Charles Darwin, as the name for the study and control of human heredity in 1883, almost two decades before "genetics" itself received a name. In an important sense, eugenics preceded genetics, and at the turn of the twentieth century it was quite difficult to distinguish between practitioners in either domain. While some leading geneticists, like Thomas Hunt Morgan, distanced themselves from aggressive eugenicist claims, many scientists pursued both. Demarcation between the two fields was difficult.

In the early twentieth century, eugenics transitioned from being a science to a vestigial science to a hyperpoliticized doctrine that was broadly criticized as pseudoscience. The basic claims of eugenics were clear enough: with the understanding of heredity (due to Mendel) and natural selection (due to Darwin), it would theoretically be possible to "improve the stock" of humanity by consciously selecting who should produce offspring for the next generation and who should be prevented from breeding. The former stance was called *positive eugenics* and included efforts to encourage the "fit" to marry and have large families (for example through subsidies or better prenatal care); the latter, *negative eugenics*, sought to deter the unfit from reproducing, sometimes through voluntary or involuntary sterilization.

Setting aside for a moment the human cost, several of eugenicists' fundamental assumptions were soon found to be problematic on technical grounds. Many of the deleterious conditions that they assumed were hereditary turned out not to be, like consumption (tuberculosis), or not to exist at all, like thalassophilia (love of the sea). Of course, some diseases *are* hereditary, but very few, such as Huntington's chorea, are transmitted by a single autosomal dominant gene and therefore could be eradicated by preventing carriers of that gene from having offspring. The Hardy-Weinberg principle (1908) demonstrated mathematically that altering the frequency of traits using eugenic measures, although possible, often required impracticable measures that slid into the morally noxious. Nonetheless, advocates of these measures continued to enact them through legislation, even receiving an imprimatur for involuntary sterilization of the "feebleminded" from the U.S. Supreme Court in *Buck v. Bell* (1927).

Eugenics was always a politicized science, but it became hyperpoliticized at the state and local levels in the United States after World War I. (And not just in the United States: there were eugenics societies and legislation in at least thirty countries around the globe during this period.) Concerns about immigration altering the demographic composition of the United States, class bias against impoverished whites, and perennial racism directed toward African Americans—blending with the claims of "racial science" about the intrinsic biological hierarchies among human races, which reached its heyday in the middle of the nineteenth century—continued to suggest eugenics as a scientific solution to what was fundamentally a political dispute. Between 1909 and 1963 (when the law was repealed), the state of California alone carried out twenty thousand forced sterilizations for eugenic reasons.

Even after the scientific community had largely walked away from eugenics, it continued as a legal force. A combination of revulsion against Nazi atrocities during the Second World War and the growing power of the civil rights movement eventually sparked a wave of reform. Rebranding continued apace, with the American Eugenics Society—founded in 1926, after eugenics had lost support within the mainstream scientific community—changing its name to the Society for Social Biology in 1972. It no longer endorsed the positions of its predecessor organization and mostly supported demographic research. (In 2014, it renamed itself once again the Society for Biodemography and Social Biology.) Yet elements of eugenic thinking remain widespread in popular culture, as a perusal of journalistic writings on genomics easily shows. Delinked from the machinery of the state, however, these doctrines simply occupy another niche on the fringe.

Chapter 4
Fighting "establishment" science

Nobody wants to be called a pseudoscientist, and so the obvious question arises: who has the authority to label someone this way? Since the issue is what counts as a science, quite often the group bestowing the label is what we have called "mainstream scientists": members of the scientific community working close to consensus methods and approaches. This is in itself a diverse group that thrives on internal disagreement and debate; sometimes, however, a subset of mainstream scientists finds a doctrine not just wrong, but also threatening. It is in moments like this that we find ascriptions of "pseudoscience."

To understand how pseudosciences get designated as such, you cannot simply start with a list of doctrines and analyze their common properties; rather, you also need to look at the mainstream scientific community and why it finds these particular ideas significant enough to attack. Plenty of mistaken—even wildly mistaken—notions get floated every day, but most of them sink without a trace. But some persist.

Those who have been labeled pseudoscientists have themselves noticed this pattern. The majority of these individuals believe they are simply doing science, but they recognize that their science is outside the mainstream consensus. They persevere because they believe this consensus is wrong, propped up by false assumptions, pecuniary self-interest, vanity, and other decidedly nonepistemological vices. They insist that in fact they themselves are the ones doing "real science," and the advocates of the mainstream are deluded. In their writings, they frequently invoke the mainstream scientific consensus as the *establishment*. This is a slur, and it paints a picture of two opposing camps: the establishment suppressing virtuous seekers of the truth, denigrating the latter with specious arguments and insults such as "crank" or "crackpot."

This is a powerful narrative, and we should not be surprised that sometimes those dubbed pseudoscientists embrace the calumnies. In doing so, they are able to draw on one of the most powerful myths underlying modern science, one which dates back to the seventeenth century: the trial of Galileo Galilei. While serving as court astronomer in Florence, Galileo publicly advocated the Copernican world system: the view that Earth orbited the Sun rather than the other way around. The Catholic Church considered this doctrine problematic, and in 1616, the Inquisition ordered him to defend it only as a hypothesis. In several works written over the next fifteen years, Galileo continued to flirt with heliocentrism, and in 1632, he published his *Dialogue Concerning the Two Chief World Systems*. The book is structured as a debate, and ostensibly Galileo treated Copernicanism as only a hypothesis, but the Church (and most readers then and now) thought otherwise. Finding him in violation of the 1616 decree, the Church demanded a recantation and condemned him to house arrest. Galileo was persecuted by the establishment, going against the mainstream consensus of natural philosophy; nonetheless his views eventually won out. This story, heavily romanticized, became a staple of scientists' self-understanding of the need to defend the truth, and it has been easily adapted for those arguing against orthodoxies of any kind—even if that orthodoxy is the scientific consensus itself.

We now come to another category of so-called pseudosciences: "counterestablishment" sciences. These are not simply anti-establishment, although that is sometimes how mainstream scientists present them, and they are not anti-science. Rather, their adherents believe that the establishment is corrupting or blocking the truth, and therefore the defenders of the real science—the demonized so-called pseudoscience—need to adapt their tactics to fight the establishment. They often do so by replicating the structures of the mainstream science they castigate. Counterestablishment sciences have institutes, conferences, journals (typically peer-reviewed), and sometimes even degree programs. In form, they display all the professionalizing markers of the establishment science they decry. Because they utilize the same mechanisms as science popularizers to broadcast their message to the public and recruit new adherents, counterestablishment sciences are among the most visible fringe doctrines.

Since counterestablishment sciences by definition replicate—or imitate, or copy, or counterfeit—the establishment they oppose, they necessarily differ depending on historical context. How the establishment operated in nineteenth-century Britain was distinct from how it worked after World War II in the United States, simply because the structures for conducting mainstream science had changed in the interim. By following a half dozen of the more significant examples of counterestablishment science, one sees not just the similarities among them, but how each reflects the dominant ways science operated in its specific context.

Phrenology

Although fringe doctrines have always existed, counterestablishment sciences really emerged only in the early nineteenth century. In order for adherents of a doctrine to replicate establishment structures, the establishment must first *have* identifiable structures, and this did not happen until the turn of the nineteenth century in Western Europe. Before then (in the age of Galileo, for example), natural philosophy was typically pursued by learned amateurs who earned their living either through another profession, such as physician or lawyer, or enjoyed inherited wealth. Scholars studying nature began to professionalize at the end of the age of Enlightenment, transforming their research into an occupation. With this metamorphosis came a new identity: the *scientist*. The term was coined in 1831, called into being by the increasingly visible presence of this novel figure who earned income from research, published in recently founded scientific journals, and was a member of learned societies. It is no coincidence that it is around this moment that the term *pseudoscience* came into being in European languages. Counterestablishment science was born.

Likely the first instance was phrenology, which originated in Switzerland but found its most solid purchase in North Britain (especially Scotland). Franz Joseph Gall, a physiologist who specialized in the brain, came to his ideas in the final decades of the eighteenth century, but they reached a wide audience only after popularization by Johann Spurzheim, who attended one of Gall's public lectures in 1800. Phrenology flourished in the new century, building on Gall's principles: the brain is the organ of the mind; the brain is not homogeneous, but an aggregate of different "organs"; each organ has a special function; all other things being equal, the size of an organ is a valid proxy for the strength of the mental faculty; and, finally, as the skull ossifies in childhood, it retains in bumps on its surface the imprint of the various organs, which can then be "read" to analyze an individual's character. As proposed by Gall and Spurzheim, phrenology was a scientific hypothesis to be investigated by physicians and physiologists; its opponents, which included most members of that intended audience, were quick to call it a pseudoscience.

Despite the official opprobrium from the establishment, phrenology remained very popular—both in the sense of having

5. This pen drawing from 1806 illustrates three perspectives of a skull, labeled according to an unorthodox system of phrenology.

many adherents, and that those adherents often belonged to the nonelite classes. Those interested in educational and penal reform intensively studied phrenological doctrines, which in turn associated those ideas with radical movements to expand voting rights to the lower classes and curtail the powers of vested interests, be they aristocrats or professional physicians and scientists. An analysis of the authors of phrenological tracts, which proliferated through the expansion of cheap print, show that antiphrenologists tended to be older and of higher social status than phrenology's advocates, who themselves tended to adhere to dissenting churches (or to no religious confession at all). British utopian socialists gravitated to the enthusiasm for phrenology, further tainting the doctrine in the eyes of the establishment. (It bears noting that some of Gall's postulates, such as the separation of the brain into faculties, are now simply part of orthodox neuroscience; the bumps on the skull, not so much.)

As the battle lines were drawn, phrenologists began to resemble their opponents structurally, even as they disagreed about everything else. There were journals and pamphlets devoted to phrenology, itinerant lecturers for public improvement, textbooks and courses, and so on. Phrenology remained popular throughout the nineteenth century (and you can still buy labeled busts as a novelty item), but it lived in a parallel epistemic universe. You could study and publish on phrenology all you liked—you just did so in separate journals. Extirpating a counterestablishment science is very difficult as long as the counterestablishment has supporters and financial resources.

Most nineteenth-century fringe doctrines paled in comparison with the vigor of phrenology. (The exceptions are Mesmerism and spiritualism, which likewise explored the connections between mind and matter and triggered popular movements of their own.) This was not for lack of trying. However, as the scientific establishment itself became more entrenched, it was harder to set up a full-blown counterestablishment to compete with it. Developing alternative institutions required persistence and financing, and the latter was often hard to come by.

Creationism

The ubiquity of creationism as the exemplary pseudoscience in the West (especially in the United States) belies how difficult it was for the movement to take off. Almost every religion has a creation myth, and even in the Judeo-Christian Bible, Genesis actually contains two slightly different versions of the Adam-and-Eve story, and there have been dozens of theologically mainstream interpretations of those few pages. Forming a scientific narrative around one specific account took time, and eventually the gradually accreting infrastructure of evangelical Christianity was ready for it. The result was the most successful counterestablishment science yet.

Creationism—the idea that the nature of Earth and the diversity of life upon it can best be explained by a creator divinity—started as a vestigial science. In the largely Christian countries of Western Europe and North America, the question was not so much whether God (understood according to the Bible) had created plants, animals, and humans, but how it was done. The most contested topic concerned humanity: whether it was created only once, so the diversity of races on the planet were the result of postcreation differentiation (monogenism); or whether the individual races were created separately (polygenism). By the time Charles Darwin published *On the Origin of Species* in 1859, the consensus for divine creation was much more troubled, with important naturalists arguing for a variety of evolutionary mechanisms. These mechanisms were not necessarily opposed to a creator deity—after all, God could have created primordial life and then let natural selection go to work—but incorporation of evolution would require significant revisions to common understandings of the first pages of Genesis.

This was not, at first, much of a problem. At the turn of the twentieth century, the framework of "scientific naturalism" promulgated most fiercely by Thomas Henry Huxley—that supernatural explanations must in principle be excluded from scientific theories—was coming to be accepted as normative, which increasingly displaced discussions about creation outside the scientific community. Around 1900, the Anglophone scientific community—most of these debates took place in Great Britain, Canada, and especially the United States—encompassed a diversity of positions ranging from pure materialism (everything came from matter without any supernatural intervention), to theistic evolution (God guided the transformation of species according to a plan), to Old-Earth creationism (the biblical account of the creation of Adam was true, but took place millions of years after the creation of Earth), to the then very marginal position of Young-Earth creationism. The last held that the universe was created in six days roughly six thousand years ago.

Young-Earth creationism was obviously close to biblical literalism, which was rising in popularity among evangelical Christians, but it faced significant problems addressing the geological observations of mountain and river-valley formation, not to mention the expanding fossil record.

The individual most responsible for addressing that conceptual problem, and as a result giving the creationist community a doctrine around which they could mobilize their counterestablishment, was George McCready Price. Born in 1870, Price was raised in the Seventh-day Adventist offshoot of the much-diminished Millerite movement—which insisted that Jesus Christ would return in 1843–1844, which did not in fact happen—so he was marginal not only in his scientific views but also in his religious ones. Trained as a schoolteacher, he intuited that the main front against the establishment should not be conducted on the grounds of Darwinian biology but through the science of geology, which provided the evidence for an Earth millions or billions of years old on which natural selection could work its wonders. In his *Illogical Geology: The Weakest Point in the Evolution Theory* (1906), Price shifted attention from Adam and Eve to Noah. In his magnum opus, *The New Geology* (1923), he argued that the biblical global flood was so violent that it catastrophically created those features of Earth's surface that seemed to be of unfathomable antiquity. This work inspired William Jennings Bryan, the venerable populist politician and defense attorney in the Scopes trial, to invite Price to testify (though Bryan disagreed with the flood geology framework), but the latter was in England and unable to travel. Price returned in 1929, and, during the ensuing decade, fundamentalists began reading Price and forming creationist societies that propounded a Young Earth bearing traces of the Noachian deluge.

During the middle decades of the twentieth century, American fundamentalists shifted their strategy from purging modernism from schools and churches, such as outlawing the teaching of evolution, and instead setting up their own radio ministries, Bible institutes, and colleges. Creationist societies fit perfectly into this separatist mold. Early attempts, such as the Religion and Science Association (1936–1937), failed to catch fire, but Price continued to be active, establishing a Deluge Geology Society in 1938, rooted in Adventist circles in Los Angeles. Yet flood geology remained peripheral in religious attempts to reconcile Christianity and science until the late 1950s, when Darwinian evolution was reintroduced into many more American public schools.

Enter John C. Whitcomb, a theologian, and Henry M. Morris, a respected hydraulic engineer. They met in 1953 and were both inspired to update Price's geological framework and harmonize it with non-Adventist theology. They presented their work, *The Genesis Flood* (1961), as a scientific treatise about a Young-Earth creation event based on empirical evidence of a catastrophic global flood several thousand years ago. Rejected by the mainstream American Scientific Affiliation, Morris and a few like-minded colleagues formed the Creation Research Society in June 1963. From the beginning, this group was more focused on education and research than evangelism and politics.

In its first decade, the society's major projects were publishing the *Creation Research Society Quarterly* and working on a high school biology textbook, published in 1970 as *Biology: A Search for Order in Complexity*. In the following decade, they started to call this approach "creation science" or "scientific creationism," and stumped for equal time with "evolution science" in high school curricula. Although the society began with five of its ten founders holding PhDs in biology, it became increasingly challenging to recruit trained flood geologists. In 1970, Morris turned down an endowed chair in civil engineering at Auburn University in order to collaborate with Tim LaHaye of San Diego (coauthor of the popular *Left Behind* Christian post-apocalyptic novels) to set up an Institute for Creation Research. (It has since moved to Dallas, Texas.) The institutionalization continued apace with journals and conferences. The public strategy was to challenge Darwinism through legislation demanding "equal time" for both "creation science" and "evolution science" in science classrooms, but in the 1980s, the United States Supreme Court determined that teaching creationism in public schools would constitute a state endorsement of religion and was therefore unconstitutional.

After the equal-time strategy failed in the 1980s, creationists took a tack less based in Judeo-Christian theology called "intelligent design," sponsored by the Discovery Institute (founded in 1990) in Seattle. All the hallmarks were there: a textbook (*Of Pandas and People: The Central Question of Biological Origins*, 1989), popular manifestos, and PhDs (some in biochemistry and mathematics; others in philosophy or the social sciences) who endowed the counterestablishment with gravitas. This legal approach of de-theologizing creationism would also fail in a 2006 appellate court decision, but the counterestablishment continues, symbiotically sustained by the fortunes of the broader evangelical movement's trend toward parochial and home schooling. The journals and the institutes still largely survive and have found imitators abroad.

Cryptozoology

Cryptozoology is not one doctrine but an umbrella term for many. Strictly speaking, the adherents of these various positions do not posit anything scientifically impossible: they claim that there are certain animals on Earth today whose existence is not recognized by mainstream science. Once one of these "mythical beings" (in the words of detractors) is observed, as in the case of the de-mythed giant squid—finally observed in 2004—then in principle mainstream scientists would welcome these organisms' existence. New species are discovered all the time.

The incredulity of the establishment toward some cryptozoologists' claims, however, has prompted the development of counterestablishments to advocate for their favored creatures. Unlike creationism, which has the full complement of structures characteristic of mainstream science, the science that cryptozoology most resembles is natural history. A discipline that privileges extensive knowledge of the outdoors and credible eyewitness testimony, it uses the kinds of evidence and the modes of communication that are more characteristic of birders and hunters than laboratory scientists.

There are a host of candidate animals (the Jersey Devil, the Chupacabra, and their ilk), but by far the two most well-known disputed animals are Bigfoot (aka Sasquatch, the Abominable Snowman, Yeti) and the Loch Ness Monster (aka Nessie). Although advocates point to folklore as providing anecdotal evidence of the existence of these beings, the properly cryptozoological quest for each dates to the early twentieth century.

The story of Bigfoot starts in the Himalayas, where a few scattered reports from British colonial officials in the nineteenth century whetted the public appetite for finding a large ape in the snowdrifts. As exploration of this region increased in the 1940s, reports of anomalous footprints and partial sightings grew, despite a concerted effort at debunking by Sir Edmund Hillary, hailed as the first person to climb Mount Everest. The locus of attention then shifted to the forests of North America. Every time an observer announced a Bigfoot/Sasquatch sighting, mainstream naturalists immediately countered that the data were internally contradictory, fraudulent, or inconsistent with the rest of natural-historical findings. Waves of interest washed over the American public about once a decade during the Cold War years, building on wistfulness about a vanishing wilderness and compromised masculinity. A major issue confronting the advocates of ABSMery—an attempt to name a scientific discipline after an abbreviation for the Abominable Snowman—has been pranksters' persistent hoaxing and the ensuing media attention. In this sense, the blending of amateur and professional in certain observational field sciences has proved problematic for erecting a counterestablishment.

Nessie provides a British inflection on the same problem: public enthusiasm, sparked in part by the opening of roads near Loch Ness in the Scottish Highlands in the early twentieth century that exposed larger stretches of the water to observation, and repeated hoaxing. Unlike with Bigfoot, where hoaxes seemed often in the spirit of fun, Nessie snapshots were occasionally monetized or hawked to encourage tourism. Scientists and granting agencies have mostly dismissed earnest requests to have the lake exposed to sonar or submarine exploration because of the expense and low likelihood of finding what was claimed to be a surviving plesiosaur (extinct since the Cretaceous) or some other sea serpent. The specific location endows these counterestablishment research groups with a municipal booster character alongside efforts to circulate newsletters and consolidate research findings.

Cosmic catastrophism

In April 1950, a Russian émigré psychoanalyst named Immanuel Velikovsky living in New York City published a bestseller named *Worlds in Collision* with Macmillan Press, the leading scientific publisher in the United States. Velikovsky claimed that a proper decoding of mythological stories drawn from around the globe demonstrated that around 1500 BCE a comet was ejected from the planet Jupiter and approached very close to Earth. This event was catastrophic to our planet, tilting its axis, fracturing the crust, and terrifying the survivors of the cataclysm, who subsequently recorded their fears in folklore. After decades of gravitational and electromagnetic interaction with Earth, the comet eventually stabilized into Venus, our nearest planetary neighbor. Velikovsky's cosmic catastrophism required the contravention of mainstream and well-established theories of geology, astrophysics, electromagnetism, and—because it relied on a controversial redating of Egyptian, Hebrew, and Mesopotamian texts so that they all described the same sensational disasters—ancient history. The book sold like hotcakes.

Part of the reason was that it offered a reconciliation of science and religion, but not on the Christian fundamentalist terms of creationism. (Velikovsky was Jewish and a passionate Zionist; he invoked no supernatural forces in his framework.) However, the real impetus for his celebrity was that a small group of scientists, including the very popular Harvard astronomer Harlow Shapley, wrote to Macmillan and demanded that they abandon the text as pseudoscientific; failure to do so might result in a boycott of Macmillan's textbooks division, the basis of its financial stability. After about a month of controversy, Macmillan acceded, transferring Velikovsky's contract to Doubleday, a press with no soft underbelly in textbooks. Velikovsky was outraged, but his sales skyrocketed. He and his partisans quickly adopted the mantle of Galileo, claiming they were persecuted by orthodox scientists for speaking the truth. The initial furor died down, and Velikovsky continued to publish companion volumes to *Worlds in Collision* while seeking legitimacy and confirmation from the scientific establishment. None was forthcoming, though he did score some successes in predicting unusual properties of Venus and Jupiter that were discovered during the incipient Space Age.

In the 1960s, cosmic catastrophism came back with a vengeance—and with a counterestablishment. As he approached seventy years of age, Velikovsky found himself swept up by the student counterculture, who imbibed his cocktail of ancient texts and interstellar science as a heady tonic. A few journals devoted to his theories appeared, while his partisans trolled mainstream scientists—especially targeting the popular astronomer Carl Sagan, who publicly ridiculed the Venus claims—demanding that Velikovsky be given a hearing. Courses about his ideas cropped up, and he lectured on campuses across the United States to packed audiences. His partisans believed that they had a new science on their hands.

Velikovsky's counterestablishment was highly personalized. He was its center and he frequently intervened in the activities of his followers, policing claims they made on his behalf and purging those he felt were not sufficiently loyal to his heterodox doctrines. When he died in 1979, the American and British offshoots of cosmic catastrophism found it impossible to maintain their momentum in the face of disconfirming geological and astronomical evidence without the force of Velikovsky's personality behind them. Within a decade of his death, the author of *Worlds in Collision* had faded into oblivion.

Aliens, present and past

Speculation about alien civilizations has always been fringy, having been a staple of science fiction from its inception, but it assumed a new cast in the postwar United States. On June 14, 1947, a foreman named William Brazel working on a ranch thirty miles north of Roswell, New Mexico, saw some unusual debris on the ground. Ten days later, Kenneth Arnold, who owned a fire control supply company in Boise, Idaho, observed nine circular objects in a chain while he was flying his airplane over the Cascade Mountains. Similar sightings of "unidentified flying objects" (UFOs) or their traces—including wreckage and scorching on the ground—sparked fascination among many enthusiasts, an interest only heightened by what seemed like ham-handed efforts of the United States Air Force to deny and cover up the traces. Over the next several decades, a counterestablishment science emerged that called itself UFOlogy.

The cinematic and literary qualities of UFOs—combined with increasing suspicion of military secrecy during the Vietnam War and the high peak of tensions with the Soviet Union—have ensured a lively curiosity about them. Sociologists, historians, and journalists have chronicled the various phases of those movements convinced that intelligent aliens have visited Earth, are still present here, and possibly have abducted (or continue to abduct) humans for experimentation. The omnipresence of cameras embedded in mobile phones and sophistication about the doctoring of digital images have somewhat dampened credulity in recent decades.

If aliens are not visiting us today, might they have done so in millennia past? In 1968, a young Swiss hotel manager named Erich von Däniken published an astoundingly successful book entitled *Chariots of the Gods?*, which explained the wonders of ancient civilizations (such as pyramids) and the universal practice of treating the gods as coming from the heavens as the result of the visitation of aliens in antiquity. They used their advanced technology to erect structures and even inbred with our simian evolutionary ancestors to produce modern humanity. The book is a fun read, combining (like Velikovsky) popular interests in both the Space Age and ancient myth. Archaeologists and anthropologists decried it as wildly distorting the evidentiary record as well as racist by implying that the dark-skinned ancients could not have accomplished these wonders without borderline supernatural assistance. The critiques did not harm von Däniken's popularity, but they perhaps diminished his capacity to erect a counterestablishment. His theories live on now mostly in popular movies and television shows.

There are two points about the counterestablishments of UFOlogy (contemporary or ancient) that bear emphasis. The first is the salience of conspiracy theories at their core. Essentially, every assertion of alien visitation comes with a corresponding claim of government cover-up. Almost all counterestablishments have an element of conspiracy-theorizing to them, summoned by the oppositional stance against the establishment of mainstream science, but in UFOlogy it is especially pronounced.

Second, on the other edge of the fringe, UFOlogy has to contend with a marginal scientific domain—the Search for Extraterrestrial Intelligence (SETI)—that stands in closer harmony with mainstream science, though most scientists overlook it as inconsequential or distracting. SETI practitioners scan radio waves bombarding Earth from the cosmos in order to uncover what might be signals from alien civilizations. The techniques here are nonconspiratorial and entirely orthodox: basic electronics, computational analysis of waveforms, calculations of probabilities. Nonetheless, the proximity to the hot-button fringe marker of "aliens!" has often endowed SETI with a tinge of the illicit, and therefore these astronomers take special pains to debunk claims of alien visitation. Reciprocally, UFOlogists smear SETI as part of the cover-up. Counterestablishments mirror establishments; sometimes the mirroring goes both ways.

Flat Earth

Although these counterestablishment doctrines continue to have adherents, hardly any of them are growing. Flat Earth theories, by contrast, have emerged from the nether regions of our culture's collective consciousness to a certain prominence. One survey in 2018 suggested that one in six Americans is not entirely convinced of Earth's sphericity, and another in 2019 registered 7 percent of Brazilian adults as hostile to the notion. There is an irony here. In popular parlance, thinking that "Earth is flat" is supposed to be a throwback to medievalism, a rejection of everything modern. In fact, since at the very latest the days of Plato and Aristotle, the Western tradition has been fully committed to a spherical Earth. The big debate was not the shape of the globe, but whether the southern latitudes were inhabited (or even habitable). With very few exceptions—such as the Christian holy men Lactantius and Cosmas Indicopleustes, writing between the third and sixth centuries—all thinkers about nature in the Christian and Islamic Middle Ages disparaged the notion of a flat Earth as nonsense. The idea that "the medievals" thought the planet was flat stems largely from the writings of the nineteenth-century American author Washington Irving, who proposed it in order to give Christopher Columbus the aura of a scientific revolutionary in heading West across the Atlantic. The twenty-first-century "revival" of this idea is not a revival at all, but a (post)modern invention.

As the recent documentary *Behind the Curve* (dir. Daniel J. Clark, 2018) chronicles, there is a growing interest in the idea that Earth is flat, with a geography centered on the North Pole and surrounded by an ice wall at the edge. The movie expertly depicts disagreements within the movement, as well as its first convention, which represents the beginning stages of a counterestablishment. These ideas build, largely without acknowledgement, on those of Wilbur Glen Voliva, who in 1914 argued that Earth was flat and the Sun was thirty-two miles across and located just three thousand miles away. (He was reacting in turn to hollow-Earth theories of the previous century's fringe, which required a spherical Earth.) Alternative geographies, such as belief in the lost continents of Atlantis and Lemuria, have long been a staple of the fringe. (And even not so fringe; Plato wrote about Atlantis.)

Even more than UFOlogy, conspiracy-theoretic thinking is very prominent among flat-Earthers. How, otherwise, to explain the overwhelming consensus that Earth is round, taught in literally every classroom in the world? And how also to account for the many images from international space missions depicting a spherical Earth hanging in space? This has to be a conspiracy to keep the masses ignorant. Indeed, this is the chief knowledge claim of flat-Earthers. There is no research profile for the group other than debunking the spherical Earth and exposing the conspiracy.

Counterestablishment movements have an additional commonality: they are heavily (though not exclusively) male. As an empirical fact, men gravitate to these fringe movements with

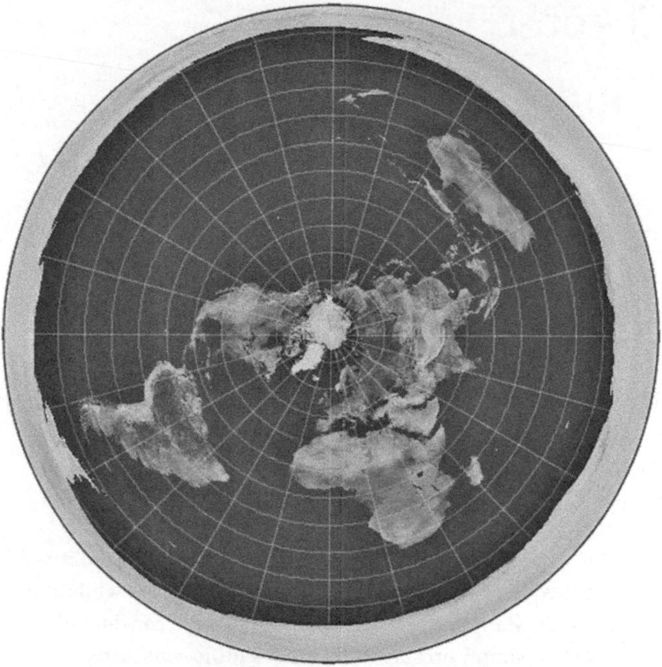

6. **This rendering of the flat Earth model depicts a central North Pole and an Antarctic outer rim, which has emerged as the consensus, though there are substantial disagreements within the movement about details.**

greater frequency than women. This can be partially attributed to the gendered domains of society they draw from—in the case of Bigfoot, hunters and woodsmen—and it is also a broader characteristic among conspiracy theorists. But at least part of it might be due to the imitative character of counterestablishment science. These institutions are built around an image of how science is conducted, and until quite recently mainstream science strongly marginalized women. Is it so surprising to see this aspect also reflected in the mirrors of the fringe?

Chapter 5
Mind over matter

The workings of the mind and its associated organ, the human brain, are terrifically complex, the motivation for the flourishing discipline of neuroscience as well as many well-established subfields of psychology. It seems that with each passing month, scientists learn something unexpected about human mental capacities, ranging from the plasticity of neurons to the abilities of a fetus to acquire the rhythms of its mother's language while still in the womb. Many of these current findings were considered outlandish or simply erroneous by the scientific consensus not very long ago.

Ask yourself honestly: Can you move objects with your thoughts? Read other people's minds? Have you ever had an eerie feeling that some major event had taken place far away—a natural disaster, the death of a loved one—and found out later that your intuition was correct? Most people would rule out the first and largely dismiss the second. The third, however, has a persistent hold. The distinguished and sober-minded American psychologist and philosopher William James, who extensively documented and experimented on claims of unusual powers of mind—what came to be called "parapsychology"—considered the anecdotal evidence of sensing a loved one's death to be so overwhelming that it merited further study, even as he dismissed other assertions as bunk.

Scientific investigations of parapsychological phenomena, especially "extrasensory perception" (ESP)—powers of mind that extend beyond the canonically recognized five senses of sight, hearing, smell, taste, and touch—demand close scrutiny. There are three reasons to single out this perpetually controversial area of research. The first is that this remains one of the most widely known of the commonly designated "pseudosciences" among both laypeople and scientists. Articles purporting to demonstrate ESP appear occasionally in professional scientific journals, written by credentialed and established scientists. (Such demonstrations remain more common in the literature than refutations of ESP. It is more exciting for editors to publish purported evidence of the extraordinary than the humdrum confirmation that telepathy or clairvoyance does not exist.) The mainstream's grudging toleration of this field grants it a special status on the fringe.

Second, there is a historical development linking the different theoretical approaches to unusual powers of mind that reaches back to at least the late eighteenth century. Even though there are sharp differences among Mesmerism and spiritualism and present-day ESP research, there are undoubted continuities of personnel and practices. (The older doctrines, as one might expect, persist as vestigial sciences alongside their more up-to-date descendants.) In this sense, parapsychology exhibits a historical evolution and differentiation that strongly resemble those of more conventional sciences.

Finally, and related to the first two, parapsychological findings have had a profound impact on the methodology of experiment that has reshaped mainstream research. The popularity and outlandish claims of past parapsychologists have always summoned a cadre of debunkers who seek to prove that the ostensible phenomena are chimeras of imagination or artifacts of experimental design. Both defenders and debunkers have along the way bequeathed to us many hallmarks of experimental practice. This methodological arms race has pushed psychology to ever greater sophistication in both the laboratory and in data analysis.

Mesmerism

The story of the mesmeric sensation sparked in Vienna, but really only caught fire in Paris. In February 1778, the Swabian physician Franz Anton Mesmer arrived at the capital of the Enlightenment from that of the Habsburg Empire, where his theories and therapies had generated considerable excitement. The boundary between medicine and physics is central here: on the one hand, Mesmer offered treatments that were supposed to palliate a number of maladies, both mental and physiological; on the other, he posited a naturalistic explanation for why his treatments succeeded. His patients often claimed recovery or at least improvement after sessions with him, and it is hard to gainsay the subjective testimony of an invalid about his or her own experience. Yet the proposed mechanism of those cures enraged French *savants*, who took it upon themselves to debunk the interloper.

Mesmer claimed to have discovered a superfine fluid that penetrated all bodies in the universe; when this fluid was blocked, sickness was the natural result. Fortunately, it was possible to control its flow through the human body and bring sufferers back to health. The fluid was magnetic, he said, and entered the body through various "poles"—the north pole channeled mesmeric fluid from the stars through one's head, the south brought up terrestrial magnetism—which could be unblocked through massaging. The rush of the fluid could generate convulsions or trances, after which the patient reported feeling significantly better. Given the curative powers of this substance, it made sense to store it for easy release, and this was possible in a huge iron tub with rods sticking out of it. The whole business (and it was a lucrative one) became tremendously compelling among both elites and plebs. It was quite possibly the most popular subject for conversation of the decade until the convening of the Estates General in 1789.

7. Dr. Mesmer often delivered animal magnetism to his adherents in crowded gatherings such as this one. Notice the large tub with metal wires in the center, which served to concentrate the fluid, as well as the elite clientele of both sexes.

That Mesmerism was sensational was not surprising. The 1780s in *ancien régime* Paris was a heady time, saturated with enthusiasm for amazing scientific findings, some imagined, some real. Shoes advertised to enable people to walk on water were fictional, on the one hand, but on the other, the Montgolfier brothers publicly demonstrated their hot air balloon in the summer of 1783, initiating human flight. For decades traveling lecturers had displayed the wonders of static electricity, charging up small boys suspended by threads and showing how pieces of paper floated up to their bodies in defiance of gravity, or discharging a shock through a chain of individuals. The ability of the Leyden jar—a glass vessel studded with a metal rod—to store electricity made it possible to see electrical effects on an even larger scale, surpassed only by the invention of the voltaic battery in 1800. Mesmerism fit into this culture of spectacle and obviously drew on the daisy-chain discharges and Leyden jar in its own iconography. Was Mesmerism like the shoes or like the balloon?

The natural philosophers of the day, shocked by reports of risqué behavior at Mesmer's salons and outraged by his speculations about a universal magnetic fluid that was undetectable by any of their instruments, aimed to resolve the question. Mesmer had asked the Royal Society of Medicine to verify his cures (it demurred), and although he was indeed invited to present before the Academy of Sciences, he was subsequently ignored by their luminaries. As he grew more popular, several academicians decided to investigate. In spring 1784, they impaneled two commissions, the more illustrious drawn from the ranks of the Academy of Sciences and the Faculty of Medicine. This was truly a star-studded crew, chaired by Benjamin Franklin, then visiting Paris on a diplomatic mission from the newly independent United States, and Antoine Lavoisier, discoverer of oxygen and the leading chemist of Europe. It was one of the earliest organizations of debunkers of the modern era, initiating a tradition in parapsychology that continues to the present.

Mesmer's aide Charles Deslon arranged to be the main subject of the investigation (over his master's protests). The commission began with the hypothesis that the magnetic fluid did not exist, and that all its subjective effects were the fruits of overactive imaginations. They set out to demonstrate this, factoring in contemporary theories of how the imagination varied by sex and social class. Lavoisier divided the experimental protocol into two parts: first, one would try to isolate the fluid from the imagination by mesmerizing subjects without their knowledge; and then one would isolate the imagination from the fluid by inducing people to believe they had been magnetized when they really had not been. The commission designed an experiment featuring a subject sitting in a chair before a screen while Deslon attempted to mesmerize her from behind without her knowledge, and then later she was told Deslon was there when he was not. The commissioners undertook a similar process with blindfolds. The results were as the commission had foreseen: no convulsions in the first instance, abundant ones in the second. This is the origin of the "placebo sham," now a standard experimental design in psychology.

The commission's published results discredited Mesmer in some circles. Others continued to believe in the doctrine, however, accusing the commission of bias and flawed methods or decrying Deslon as a poor magnetizer. Then the French Revolution swept Paris and Mesmerism was displaced from public consciousness. It found a new home across the English Channel in the 1820s. During the British revival of Mesmerism, adepts used magnetic fluid to anesthetize subjects before surgery and combined magnetism with phrenology as a new conduit to understanding the brain. Once again debunkers sallied forth, attacking those physicians who swore that, whatever its scandalous origins, animal magnetism was a plausible scientific phenomenon.

Spiritualism

"Spiritualism" (or "mediumism") was invented in upstate New York in 1848 but blossomed in Victorian England, where it hybridized with the remnants of Mesmerism. In Buffalo, the Fox sisters claimed to be able to enter a trance state and communicate with departed spirits, who responded by rapping on tables, moving furniture, and the like. In London, these impromptu American séances acquired a more formal structure.
A "medium"—so called for mediating between the spirit world and ours—typically a young woman or an underage male from the lower classes, would sit with a group around a table in a darkened room. According to most practitioners, disembodied spirits would flow through the medium and register a variety of effects: knocks and raps, levitation of furniture or people, automatic writing, spooky sounds, and occasional ectoplasmic manifestations. Spiritualism too became the talk of fashionable circles, and

offshoot movements sprang up in Paris, Berlin, St. Petersburg, and farther afield.

Many scholars who have analyzed the spiritualism movements of the late nineteenth century understand them as a manifestation of the crisis in traditional religious belief. Given the tremendous advances of science of the day, members of the midcentury middle class worried that the ecclesiastical rituals they had inherited smacked more of superstition than of truth. In this sense, spiritualism was for some a more scientific way of understanding the soul: it relied on observation in controlled spaces and often used new instruments—such as the emergent technology of photography—to document what happened during séances. Séances were fundamentally empirical, since participants observed what was happening and often made strenuous efforts to detect fraud or hoaxing; if the medium survived these trials, why withhold assent to the reality of the phenomena? By this interpretation, spiritualism was neither antiscientific nor antireligious, but an amalgamation of both science and religion.

Emphasizing too much of the "spiritual" in spiritualism risks flattening the movement's heterogeneity. Some of those who participated in séances remained agnostic (the term itself a coinage of this era) as to whether departed souls caused the phenomena in the séance room. Perhaps these effects were physical products of interactions between impulses in the brain and the electromagnetic ether? Many highly respected scientists approached séances with a skeptical eye before becoming convinced of their own observations and lending mediums their valued scientific authority. These included leading British chemist William Crookes, the co-discoverer of natural selection Alfred Russel Wallace, and the highly influential Russian organic chemist Aleksandr Butlerov. A cohort of British men (and some women) of science assembled in 1882 into the Society for Psychical Research

(SPR), under the presidency of philosopher Henry Sidgwick. (The American Society for Psychical Research was founded two years later.) Although most of the early members joined the SPR because of their interest in spiritualism, the group expanded its inquiries into telepathy and hypnosis, which were understood as related phenomena. Meanwhile, spiritualist circles splintered into mystical movements such as theosophy, which blended in esoteric South Asian thought.

As with Mesmerism, spiritualism attracted plenty of debunkers, including the nascent profession of performing magicians such as Harry Houdini, who used their skills to unmask frauds. Some mediums survived the onslaught and continued to draw adherents. Scientists were most concerned that the prominence of their peers among the believers might lend more credibility to the movement, and several investigatory commissions were launched. In St. Petersburg, for example, Dmitrii Mendeleev, who had formulated the periodic system of chemical elements in 1869, organized one that fractured his relationship with Butlerov without stemming the tide of interest in the occult (notwithstanding its damning findings).

An interesting product of these investigations—both under the auspices of the SPR and outside of it—was the introduction of randomization into experimental trials. Many of the tests of mediumistic powers hinged on the ability of these individuals to guess sequences of playing cards or the identity of other hidden objects with surprising accuracy. Parisian physiologist (and later Nobelist) Charles Richet was inspired by reading the SPR's journal to test whether telepathy might be more generally distributed in the population, and, in the 1880s, he introduced randomized groups of subjects and cards to screen out lucky coincidences or trickery. This innovation soon migrated from the murky domains of parapsychology to become perhaps the most important change in experimental practice of the past two centuries.

University parapsychology

At the turn of the twentieth century, psychical research was conducted in an amateur manner, whether by those without professional credentials or by trained scientists who entertained it as a sideline without substantial resources (or the approval of their colleagues). In 1898, Harlow Gale, an on-again-off-again instructor in psychology at the University of Minnesota, polled psychologists at eleven institutions about the state of psychical research, only to find that just two were pursuing it seriously (himself and the aforementioned William James at Harvard). They were soon joined by J. H. Hyslop at Columbia University and an unnamed lecturer at the University of Pennsylvania. When Hyslop repeated Gale's enquiries in 1917, he found that academic psychical research had deteriorated even further. When universities were left to their own devices, they discouraged such work.

Wealthy donors had their own ideas. Even as Hyslop conducted his second census, bequests specifically for psychical research showered on Clark University (1907), Stanford (1911), and Harvard (1911–1912). Thomas Welton Stanford, younger brother of Stanford University's founder, donated $50,000 for a division of psychical research in the psychology department. Both the department chair and the university's president accepted the gift with misgivings. They appointed John E. Coover, a homegrown recent PhD, to do this research. Coover was skeptical, yet he devoted five years to rigorous experiments, in the process developing a model for how to apply experimental psychology to this area. He published *Experiments in Psychical Research* in 1917 as the endowment expired, attributing the results to nothing but chance, and also exposed a fraudulent medium. Harvard's fund in memory of Australian-English psychical researcher Richard Hodgson was left fallow until administrators finally found a suitable recipient in 1916–1917. Both administrative disdain and scarce personnel offered a grim prognosis for psychical research.

Nonetheless, interest in moving the study of the paranormal into the professional ranks deepened after the First World War, as new cohorts of no-nonsense experimentalists began to join the American Society for Psychical Research. By the early 1930s, a change of generations had taken place, enabling the introduction of new methods drawn from Ina Jephson's pioneering experiments on clairvoyance at the British SPR in the 1920s. This revitalized tradition of parapsychological research found its champion at Duke University in the person of Joseph Banks Rhine, who shifted from botany to psychology when he joined the North Carolina university in 1927.

Parapsychology as it developed professionally in the twentieth century (and to a certain extent as it exists today) crystallized around Rhine's program at Duke, and especially his paradigmatic book, *Extra-Sensory Perception* (1934). Although he had long been interested in psychical investigations, he had early on been disappointed by fraudulent mediums and dispirited by SPR-style work. He launched his program in earnest in 1930 after reading extensively in the published literature and recruiting experimentalist colleagues, such as perceptual psychologist Karl Zener.

The refinement of methodology distinguished Rhine's research. Earlier work, dating back to before Richet, had used the capacity of certain individuals to guess the faces of concealed playing cards. Rhine appreciated the simplicity of the arrangement and the ease with which the results could be subjected to statistical analysis, but he found it beset by too many confounding variables. Individuals might have particular emotional associations with specific playing cards (say the ace of spades or the queen of hearts), which might bias their guesses, and there were simply too many different cards to work with. Instead, he commissioned what are now called Zener cards, which removed the question of suit and replaced the numbers with five black shapes: a circle, a plus sign, three squiggly waves, a square, and a five-pointed star.

A subject guessing at random ought to get a rate of 20 percent correct (one in five); anyone significantly above chance—or, intriguingly, significantly below—in a statistical sense would provide evidence for some psychical abilities. Following Jephson, Rhine conceptually distinguished between two effects: telepathy (reading the experimenter's mind as the latter looked at the Zener card) and clairvoyance (identifying the card when it was hidden from both experimenter and subject). Subjects went through thousands of card runs in grueling shifts. Most ended up around chance, within experimental error; some, however, did not. By bringing new techniques of statistical analysis and experimental controls into psychical research, Rhine raised both the respectability of the field across the academic world and donations for Duke.

Many of the skeptics' criticisms were unsurprising. Instead of elaborating a causal mechanism—how the images were transmitted to the percipient—Rhine stressed the primacy of the statistically significant effect as demonstrating that the phenomenon existed, even if he could not explain how it worked. That his ESP results did not weaken with distance, so that gifted subjects guessed correctly at the same rate whether the cards were in the next room or in a different city, seemed to him a problem for future researchers to chew on. Irving Langmuir developed his demarcation criterion of "pathological science" in explicit reaction to Rhine's results. Even the most statistically minded scientists, such as population geneticist George Price, were perturbed by the implications of the data, and insisted that the only plausible explanation was fraud, a charge that stung the punctilious Rhine.

Indeed, crusading skeptic and debunker Martin Gardner would note in 1957: "It should be stated immediately that Rhine is clearly not a pseudo-scientist to a degree even remotely comparable to that of most of the men discussed in this book. He is an intensely sincere man, whose work has been undertaken with a care and competence that cannot be dismissed easily, and which deserves

8. Joseph Banks Rhine conducts a parapsychology experiment using Zener cards.

far more serious treatment than this cursory study permits." By 1940, about fifty universities were experimenting with Zener cards. Yet, enthusiasm for Rhine's work tapered off in the 1950s. It was not that psychologists dismissed the results per se, but just that they had limited energy and resources and preferred not to work on a topic that so clearly incensed the leadership of the American Psychological Association.

University-based parapsychology did not go extinct, though it never again attained the level of mainstream credibility it had earned with Rhine. In the mid-1970s, about half a dozen institutions of long standing had a smattering of full-time parapsychology researchers. In 1969, the Parapsychology Association was officially admitted to the American Association for the Advancement of Science, though most publications in this area continued to appear in specialized (and marginalized) journals. A developing area in the 1970s promoted mechanical systems that were known to be random—rolling sixes abnormally frequently on a fair die, for example—testing subjects' psychokinetic potential to manipulate matter.

One of the last such programs was the Princeton Engineering Anomalies Research (PEAR) Laboratory, located in the School of Engineering and Applied Sciences of the Ivy League university in New Jersey. It was the personal project of Robert Jahn, a leader in designing electric propulsion for spacecraft and dean of the school, who was reportedly turned on to researching the topic by an undergraduate student. The PEAR Lab built random-number generators and then tested subjects' ability to skew the average either above or below the statistically expected mean output. Similar to Rhine, Jahn published data that indicated an effect (at a significance level of 0.00025), but, in contrast to Rhine's work, researchers at other laboratories were unable to replicate them. The PEAR Lab continued on private donations, in particular seed money from aerospace mogul James McDonnell (who believed that the mental states of pilots influenced aircraft machinery), but it closed in 2007 and moved off campus, to the relief of his former engineering colleagues.

Debunkers

Most of the opposition to parapsychology in the twentieth century stemmed from genteel disregard by mainstream psychologists for what they saw as an embarrassment to a social science that was trying to gain respectability among natural scientists. It was largely passive. With the 1970s resurgence of interest in many forms of heterodox science, some skeptics took on a more aggressive stance, assembling a motley cohort to engage in systematic debunking of claims of the paranormal ranging from ESP to UFOs.

Symptomatic of their concern was the Israeli-born mentalist Uri Geller. People interested in parapsychology began to hear of reports of Geller's amazing abilities in 1969; by the early 1970s, he was traveling widely in Europe and North America, demonstrating before packed halls and on television a capacity to guess the numbers on dies hidden in boxes, read minds, and—his signature feat—hold a metal spoon in his hand and bend it with his mind. (This last was featured prominently in the 1999 science-fiction blockbuster *The Matrix*, where the spoon-bending reference was likely lost on younger viewers who did not remember Geller.) Among those captivated by the media phenomenon was Andrija Puharich, a parapsychological investigator who arranged in September 1972 for Geller to be brought to the Stanford Research Institute (SRI), a nonprofit research center that had formally separated from the university in 1970.

Two former laser physicists at SRI, Russell Targ and Harold Puthoff, conducted a barrage of studies on Geller and concluded that he was genuine. In October 1974, they published a paper based on this work in the very prestigious British scientific journal *Nature*. (This piece is one of the few parapsychological efforts that managed to land in such a prominent venue.) The work was partially funded (to the tune of $50,000) by the Central Intelligence Agency, which was worried about an emergent "psi gap" with the Soviets, who were believed to be recruiting psychics to engage in what has been called "ESPionage": the reading of files in locked cabinets using clairvoyance. NASA also provided some funding, urged on by astronaut-telepathist Edgar Mitchell, who himself had tried an ESP experiment from the surface of the Moon in 1971.

All of this infuriated the skeptics, who started to believe that the conventional mechanisms of the scientific community were unable to stop an onslaught of new irrationalism. One particular project emerged from, of all things, frustration at astrologers. Secular humanist philosopher Paul Kurtz had coauthored a prominent 1975 attack on astrology, then experiencing a bit of a revival (as it is today), and turned his attention to organizing opposition to parapsychology. He put together a group called Resources for the Scientific Evaluation of the Paranormal, and invited a diverse array of fellow skeptics including popular science writer Martin Gardner, sociologist Marcello Truzzi, and prominent magician James "The Amazing" Randi. The group was soon reconstituted as the Commission for the Scientific Investigation of Claims of the Paranormal, with the formidable acronym CSICOP (pronounced "sci-cop"; get it?). CSICOP incorporated some very heavy hitters, most prominently ubiquitous astronomer Carl Sagan, science fiction luminary Isaac Asimov, and eminent psychologist and behaviorist B. F. Skinner. Sagan, who had a reputation for tolerance of the fringe given his interest in the possibility of extraterrestrial intelligence, was one of the most visible members of the group.

CSICOP undertook some experimental investigations of parapsychology, but mostly it was interested in exposing frauds in sensational ways. Truzzi quit for precisely this reason. Officially, Randi has written, "The CSICOP does not deny that such things *may* exist, nor do I, personally. However, in light of my considerable experience in examining such matters, I will say that my assigned probability for the reality of paranormal powers approaches zero very closely." Randi continued to offer a substantial check to anyone who could do a single paranormal feat of any kind under proper observing conditions, and never had to pay out before his death in late 2020. CISCOP was controversial from the start; Geller sued it for $15 million for defamation, but the case was tossed out in 1991. The organization still exists under the name Committee for Skeptical Inquiry, and it continues to attack paranormal claims.

Debunking has not been a huge success. It is true that individual claims of the paranormal are found to be fraudulent or dissolve upon closer inquiry, but others crop up. The willingness of high-profile scientists to engage in efforts like CSICOP has also diminished over the years. Although revulsion at parapsychology remains widespread among the mainstream scientific community, decades of dedicated debunking has not deterred new contenders. Two examples indicate the distance between concerted campaigns and other mechanisms for maintaining ESP on the fringe.

The Welsh-born physicist Brian Josephson is perhaps the most prominent name in the parapsychology community at present. He is also the only one with a Nobel Prize to his name. He won that award in 1973 for the theoretical prediction of the eponymous "Josephson effect," which describes quantum tunneling through a superconducting barrier. Josephson had done his initial work on superconductivity while a twenty-two-year-old graduate student at Cambridge and by the early 1970s had found his interests shifting toward Transcendental Meditation and other fringe ideas. (He consulted with SRI's Puthoff and Targ, for example.) The Nobel freed him up to pursue these interests further, and he has remained a vocal decrier of the stifling blinders imposed by mainstream scientific orthodoxy. CSICOP and its like are no deterrent.

Finally, there is the case of Daryl Bem, now professor emeritus of psychology at Cornell University. In 2011, he published an article in the *Journal of Personality and Social Psychology*, a leading outlet in that subfield, entitled "Feeling the Future," which claimed—using Rhine-style statistics—to detect precognition. The piece was heavily attacked by the mainstream of the community and was one of the triggers for what has come to be called the "replication crisis" in psychology, the scientific community's internal critique of experimental methodology in the wake of persistent failure to replicate even canonical scientific results.

These cases—Bem, Josephson, Jahn, Rhine—indicate an important feature of modern science: its home in the research university. Of course, plenty of science also takes place in government or industrial labs, but the role of the university as a central locus for the production of science is not in question.

This is one reason why parapsychology is so illuminating: it lives on the edge of the university, straddling the border with counterestablishment sciences but not quite of them. While ESP research has taken place in universities, it seems to be unable to sustain that position for long. Demarcation eventually shows up to reinforce borders.

Chapter 6
Controversy is inevitable

The wide sweep of doctrines that have been called pseudosciences—from astrology to cryptozoology, from creationism to Aryan Physics, from parapsychology to alchemy—do not share enough of a common essence so we can declare: "Beliefs that claim to be sciences but possess properties x, y, and z are pseudosciences." That does not mean, however, that we are unable to find occasional commonalities in their histories. Though a "pseudoscience" may not be a single identifiable thing, the act of labeling it follows a fairly common process.

Pseudosciences can start out as sciences—as with the vestigial sciences of astrology, alchemy, and eugenics—and then gradually fall out of favor (typically through extensive theoretical and empirical criticism); any remaining adherents find themselves advocates of a fringe idea. Others are, in a manner of speaking, born pseudoscientific. Velikovsky's propositions about cosmic catastrophism and ancient myth, or enthusiasm for the Yeti or the Loch Ness Monster, did not begin as domains of science, but were excoriated from their first appearance by the mainstream. Yet the process of fringing is the same in both variations: it is generated by the consensus of the relevant group of scientists. When the consensus shifts decisively against an idea, and instead of abandoning it its advocates double down, there is a strong chance that their beliefs could be labeled pseudoscientific.

Yet between those two extremes (starting scientific, starting pseudoscientific) there is a gray area. Take the case of French physicist Prosper-René Blondlot. Blondlot was a respected scientist with a series of early successes in the cutting-edge field of electromagnetic radiation. In 1891, he made the first measurement of the speed of radio waves as 297,600 km/s, which happens to be within 1 percent of today's accepted value of the speed of light, forming an important experimental confirmation of the theory of electromagnetic radiation postulated by James Clerk Maxwell. In 1903, Blondlot claimed to have discovered a new kind of radiation, which he called N-rays, named by analogy with the sensational X-rays discovered by Wilhelm Röntgen in 1895 and in homage to his home city of Nancy. He measured the presence of N-rays by observing the changes in brightness of a spark. The discovery was met with broad interest and many European scientists rushed to replicate the finding (some successfully). However, a year later, an American physicist named Robert W. Wood, having visited Blondlot's lab and examined the set-up, averred that, when he surreptitiously removed a crucial part of the experimental apparatus while Blondlot was taking readings, the latter insisted he continued registering N-rays. Wood attributed the "discovery" of N-rays to Blondlot's suggestible imagination, and within a year the previous experimental findings were dismissed as artifacts. N-rays were determined never to have existed. Blondlot's reputation never recovered.

How should we understand this case? It does not seem like Velikovsky's, since Blondlot was a member of the scientific community in good standing and N-rays were treated as plausible when first announced. Then again, this was not quite like eugenics, either, given that the properties of N-rays were controversial from the outset and were subject to pointed skepticism throughout their brief heyday. It is tempting to consider this as a hallmark example of a pseudoscience; indeed, Irving Langmuir cited it as a canonical example of "pathological science." Yet before Wood's exposé, it seems like Blondlot was performing research similar to his measurements of radio waves. He was, in short, pursuing science normally.

This is a troubling claim, but it should not be a surprising one given what we have already seen. Among the doctrines classified as pseudosciences by the scientific community, there is a sizable complement of vestigial sciences, which by definition once counted as sciences and then ceased to. What makes them pseudoscientific today is that a significant group of people are still defending them as scientific after the mainstream consensus has decided otherwise. The straightforward implication is that *any* scientific position could receive the label of "pseudoscientific" depending on its future trajectory. Since we do not know the future, any present science has potential disgrace waiting in the wings. Not only is this possible, it is practically inevitable given two structural features of contemporary science.

First, today's science is adversarial. The way a scientist makes her reputation is by building on past findings, but if all she does is confirm what everyone already knew, her career stagnates. The pressures in scientific research are to do something new, and that usually means refuting a tenet of contemporary science. (We detect echoes of Karl Popper's falsificationism.) Credit in science is allocated for priority (being first) and for being more correct than your competitors investigating the same questions. There will always be winners and losers. If the losers persist, they can and will get shunted to the fringe.

The second reason is that science is increasingly expensive. There are limited resources to go around, and there are always too many researchers chasing after coveted grants and high-profile publication opportunities. Within a climate of scarcity, adversarial norms necessarily generate both an incentive for winners to defend their gains and resentment from those who lost. Anyone who jeopardizes your research—say, by defending a fringe theory that contradicts it—may be seen as a threat. When nonmainstream doctrines pose a threat (real or imagined) to professional scientists, the term *pseudoscience* gets bandied about.

Demarcation is built into our funding systems. Applicants need to present their own work as superior to those of wrong-headed competitors, and the panels that evaluate the grants must always reject a large number of proposals as less worthy than the few they endorse. Limited funds set up a ruthless machine for discarding scientific claims, some of which might end up on the fringe. Studying the category of pseudoscience thus yields some insights into how contemporary science works.

The gray area is produced by the fact that almost every significant new scientific claim can potentially be the subject of controversy, the fuel that powers the cycles of credit and reputation. But not all discarded doctrines experience the same fate. Even in a single domain—the scientific properties of water—some of the losers of controversies end up simply as yesterday's news, sincere science that happened to be mistaken, while others are branded as ignominious and take up residence on the fringes of knowledge.

Polywater

In the mid-1960s, many chemists across the leading scientific nations enjoyed the pastime of studying liquid water's fundamental properties. Water is a weird substance, scientifically speaking: it is, bizarrely, more dense as a liquid than as a solid (this is why ice floats), and it has a higher melting point and specific heat than theory would predict. It is also very easy to access and studying it does not require expensive equipment. To be sure, the study of water's properties is not typically regarded as a pressing issue, but it has sustained a lively scientific interest.

The first reports of "anomalous water" emerged from the laboratory of Nikolai Fediakin at the Technological Institute of Kostroma, a small city two hundred miles northeast of Moscow.

Fediakin was experimenting with liquids sealed inside very narrow glass capillary tubes. When he left water in the tubes over several days, he noticed a secondary column growing at the top—it seemed that the water was spontaneously fractionating into two distinct components. He published this result in 1962 in a Russian-language journal focused on colloids. At that point, a much more notable Soviet scientist, Boris V. Deriagin (also spelled Derjaguin), director of the Surface Forces Laboratory at the Institute of Physical Chemistry of the Academy of Sciences in Moscow, took over the research. His laboratory published ten significant articles on the phenomenon between 1962 and 1966.

The properties his team uncovered were shocking. This second fraction of water seemed to be chemically identical to ordinary water (that is, its molecular composition was H_2O), but otherwise it was puzzling. Compared to "regular water," its thermal expansion between 20°C and 40°C was about 1.5 times higher, and its freezing point and boiling point were lower and higher, respectively—in the case of the latter, reaching about 250°C! Stranger still, the density of this new water was 10 to 20 percent higher than normal, and its viscosity was syrupy.

The research remained confined to the Soviet Union until 1966, when Deriagin traveled to Nottingham, England, to participate in a discussion of the Faraday Society, at the invitation of J. D. Bernal, the director of the Crystallography Department at Birkbeck College, University of London, who had a long-standing interest in water. Bernal, a distinguished scientist as well as a prominent member of the British Communist Party, had stood out for his continued support of Lysenko's theories after Stalin's official endorsement of Michurinism in 1948. His Soviet connections had brought Deriagin's work to his attention. Nonetheless, broader scientific enthusiasm did not build until 1969, when an American named Ellis Lippincott announced similar astonishing results at a meeting of the American Chemical Society in New York City. Lippincott rechristened Deriagin's "anomalous water" as "polywater," under the hypothesis that it was a polymerized form of water, the molecules being linked together the way they are in plastics. The following year saw the peak of research on polywater, with Soviet scientists focusing on its bulk properties and Western researchers concentrating on its microstructure. The research received substantial support from the Office of Naval Research, a major funder of American science, and rumors circulated that Deriagin was shortlisted for a Nobel Prize. Much of the publication about polywater came in the form of short communications to *Science* and *Nature*, the most prestigious scientific journals in the world both then and now, emphasizing polywater's ascendency.

Within a few years, however, the bloom fell off the rose. By 1973, both American and Soviet researchers had published results indicating that the anomalous behavior of the polywater molecules was best explained by impurities in the water samples. The brouhaha was, in short, an artifact. A polywater advocate today would likely be labeled a pseudoscientist by the establishment, but you are unlikely to find such a person. Polywater was a textbook scientific controversy: hotly debated according to the conventions of the field and subsequently resolved and abandoned. Unusual in this case was the international character and high visibility of the debate.

Water memory

On June 30, 1988, the prominent journal *Nature* published an article entitled "Human Basophil Degranulation Triggered by Very Dilute Antiserum against IgE," hailing from a laboratory at Paris's Institut national de la santé et de la recherche médicale (INSERM), directed by immunologist Jean Benveniste. The article reported an experiment that took highly diluted solutions of a particular chemical compound, so dilute that statistically any sample to an extremely high probability contained not a single molecule besides water. Nonetheless, Benveniste and his coauthors reported, these samples still retained some properties of the substance that had initially been dissolved in them—something diluted to the level of 10^{120} could have the same effect as something diluted to 10^{2}. It was as though the water had a "memory" of the substance that had initially been stored within it.

There was something odd about this article beyond the surprising findings it contained. The editors of *Nature*, led by John Maddox, published a companion piece in the same issue that critiqued the INSERM results. Such a practice was, to say the least, highly unorthodox, but then again so were the claims Benveniste's team was making. The notion that dilute solutions can "remember" the properties of a solute without its presence was a central premise of homeopathy, the alternative—and highly controversial—fringe medical practice. Contrary to mainstream "allopathic" medicine, which combats ailments by using an opposing material, homeopaths insist on using a similar substance, such as highly dilute solutions of potent poisons like arsenic or belladonna. The similarity to homeopathy provided the element of threat that triggered Maddox's unusual actions.

He did not stop at just the companion article. From the moment the article was sent out for peer review the previous year, Maddox cooked up a plan to turn the Benveniste findings into an object lesson for how to conduct a scientific controversy. The peer-review process had been fraught, with some referees decrying the homeopathic implications; Maddox chose to publish the article anyway. He then impaneled a team to go to INSERM and investigate the investigators. Rather than a cohort of immunologists, the team consisted of Walter W. Stewart, an employee of the U.S. National Institutes of Health with a (somewhat contentious) reputation for investigating allegations of elite scientific fraud; James "The Amazing" Randi, the magician who debunked psychics; and Maddox himself. This was a hanging jury, and hang they did. In the meantime, *Nature* suspended its usual practice of publishing letters engaging with the INSERM article until the investigation had concluded.

On July 28, 1988, within a month of the initial publication, the team published its report: "'High-Dilution' Experiments a Delusion." The article argued that the statistical analysis in the initial paper was faulty and that INSERM's own double-blind experiments did not replicate the results. Stopping short of alleging deliberate fraud, the team claimed that the French researchers had been misled by their own enthusiasm, similar to the accusations Blondlot had faced. *Nature* also published the rebuttal from INSERM. The controversy—and the controversy about Maddox's handling of it—continued for another eight weeks, when Maddox shut it down.

The water memory debate seems a hybrid between the relatively ordinary scientific controversy of polywater and the debunking culture of the anti-ESP scientists, distinguished further by the unusually interventionist role of a scientific journal. It is also important that the term *pseudoscience* was not a prominent part of the handling of this affair, though the term does attach to those—like certain members of the homeopathy community—who continue to cite the INSERM *Nature* article as scientific support. The boundaries between fringe, fraud, and mistaken science are blurry.

Cold fusion

The experiment itself was elegantly straightforward. Two electrochemists at the University of Utah in Salt Lake City, chair of the chemistry department Stanley Pons and Czechoslovak-born British expatriate Martin Fleischmann, took a flask of an electrolyte dissolved in heavy water—water with some of the atoms of hydrogen replaced by its heavier isotope, deuterium, which has an extra neutron—and ran a current through it from two electrodes.

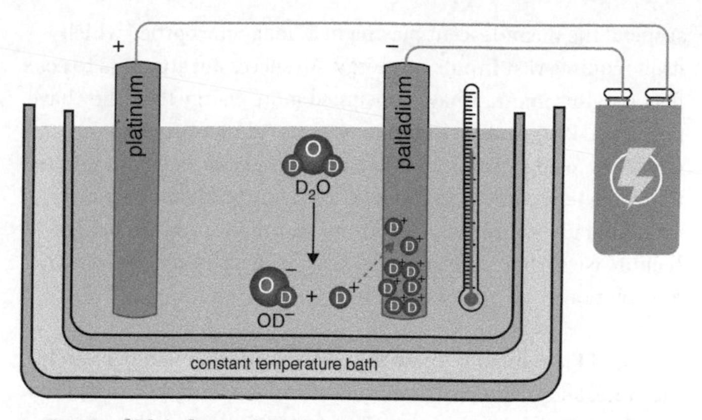

9. Pons and Fleischmann's cold fusion cell was constructed out of very simple components, the most important of which was the palladium electrode.

Less straightforward is what they said happened next. One of their electrodes was made of palladium, a metal with the interesting property of highly concentrating hydrogen ions from a solution. The electrochemists maintained that when they ran the experiment, they detected an anomalous increase in heat, as well as a flux of neutrons. The explanation for these effects, they stated, was that the nuclei of the deuterium ions packed inside the palladium electrode had fused together into helium, releasing enormous energy.

If true, this would be the most important scientific finding of the century. Hydrogen is the most abundant element in the universe, and fusion of its nuclei into heavier atoms is what powers the stars, including our Sun. Attempts to harness fusion in a controlled manner—we know how to unleash it in an uncontrolled way in thermonuclear (also known as hydrogen) bombs—have been underway since the 1950s, but face tremendous challenges. Hydrogen nuclei fuse under enormous pressure and heat, heat so extreme that it would melt any container, so current experiments suspend the incandescent plasma in a "magnetic bottle," which itself requires vast inputs of energy. All successful attempts to coax fusion in this manner have consumed more energy than they have produced. Pons and Fleischmann suggested that they had produced "cold fusion": fusion that required less energy to initiate than it generated. This could lead to energy production from an unlimited fuel source whose only waste product was the benign helium. No carbon emissions, no reactor meltdowns—it would be a revolution.

Pons and Fleischmann knew the excitement their findings would generate, and they stage-managed the announcement to heighten the effect: they held a press conference. This was a collaborative decision with the technology transfer office of the University of Utah, which was eager to generate interest in the patents (and subsequent lucre) that would eventually flow from the commercialization of cold fusion. But it was also a decision meant to forestall a competitor and vouchsafe their priority in the discovery. Physicist Steven E. Jones at nearby Brigham Young University in Provo, Utah, was working on similar questions in geophysics, such as how a kind of cold fusion of concentrated hydrogen atoms might cause the unexpected abundance of helium in certain rock formations. Jones had contacted the University of Utah group about a presentation he was going to make at a conference in March 1989, and this initiated plans for a joint announcement for later that month. When Jones canceled his attendance at the conference and instead submitted an abstract for the American Physical Society annual meeting, scheduled for May 1 in Baltimore, Pons and Fleischmann believed that Jones had reneged on their agreement. Initially, the three researchers had planned to submit their papers to *Nature* together, but now the Utah chemists submitted alone to the *Journal of Electroanalytical Chemistry*, having received a promise that it would be published on April 10. Jones was furious, and Pons and Fleischmann stepped before the cameras to make sure that they received credit for the Earth-shaking discovery.

The press conference dominated the news cycle, and Pons and Fleischmann became household names. Today, press conferences to announce groundbreaking discoveries are not especially unusual, but they still were in the late 1980s. Pons and Fleischmann displayed the basic experimental arrangement and laid out the potential implications, to the delight of reporters. Scientists were rather less pleased. Few relevant technical details were provided—ostensibly at the request of the editors of the *Journal of Electroanalytical Chemistry*, where the article was still under review—and immediately scientists across the country attempted to replicate the findings using the clues they had: video footage of the experiment traded among research groups by mailing videocassettes or faxing grainy photographs. Some early replications came in, but more common were failures. There were theoretical problems, too. To produce the amount of heat Pons and Fleischmann had reported, models of nuclear fusion predicted an emission of radiation so high it ought to have killed them both. Most scientists thought to wait until the publication was released to learn more specifics.

Pons, Fleischmann, and the lawyers at the University of Utah did not wait. They headed to the Utah statehouse and the U.S. Congress to request sizable grants in order to develop cold fusion technology at a larger scale. Meanwhile, the few replications that had been announced were retracted: a faulty neutron detector here, a miscalibrated thermometer there. At this point, cold fusion was still controversial science.

The bottom fell out on May 1, 1989, at the very American Physical Society meeting that had caused the rift between the University of Utah team and Jones. Physicists—for scientific reasons stemming from the available evidence, but also because of hostility to the showboating chemists who trod upon their domain—engaged in a high-profile debunking session, with an especially damaging takedown by Caltech chemist Nathan Lewis, that tore apart what little survived of Pons and Fleischmann's original claims.

It remains a matter of debate whether the original findings were an experimental artifact (like polywater), experimentalist overinterpretation (like water memory), or deliberate fraud. The world took notice of the physicists' onslaught. One way to measure the bubble of cold fusion euphoria is to track the price of palladium, the essential metal for the electrodes in the Pons-Fleischmann setup. In March 1989, it was trading at $145.60 an ounce. By May, just before the APS meeting, it had shot up to $170. Afterward, it collapsed to $95. Pons and Fleischmann were disgraced, and both moved to France in 1992 to continue their research, though their new laboratory closed in 1998 without achieving significant results.

Yet the death of cold fusion has been greatly exaggerated. A small group of researchers continue to explore the Pons-Fleischmann approach to energy generation. Specialty journals like *Cold Fusion* and *Infinite Energy* started up in 1994 and 1995, respectively. Every year since 1989, there have been International Conferences on Cold Fusion (since 2007 called International Conferences on Condensed Matter Nuclear Science), with the noticeable participation of Japanese researchers, who receive some state funding for this work. Even Steven Jones, uninvolved in the scientific scandal, got embroiled in a controversy of his own. He became a founding member of Scholars for 9/11 Truth, arguing that the destruction of the World Trade Center in New York City was not the result of a terrorist attack. Although he resigned his membership in 2006, Brigham Young University put him on paid leave and then negotiated his early retirement.

Fraud and the replication crisis

Controversy is inevitable. It is part of what it means to do science in its modern form. Researchers come up with new ideas and competitors try to knock them down. Sometimes the innovators succeed in establishing their doctrine as the new orthodoxy, and sometimes they fail. The question is: what happens to those who lose? Do they concede their errors and participate in the new consensus, or do they continue to maintain they had been correct? If the latter, for how long do they keep it up? If it is for long enough and the potential threat they pose to the establishment is distracting enough, they could come to be designated as "pseudoscientists" (as some, but not all, cold fusion researchers are labeled today).

The controversies that end up careening into the fringe are related to, but distinct from, two other phenomena that are increasingly visible in contemporary science. The first of these—already hinted at in the cold fusion debates—is scientific fraud. It is difficult to determine whether fraud is increasing in contemporary science, since we do not have a reliable baseline for the ubiquity of past scientific fraud; doubtless, however, it is more visible today than ever. The stakes of a successful scientific career (prestige, income) are quite high, and it is understandable that some individuals might follow those incentives so far as to fabricate or misrepresent their data in the hope of producing a higher-ranked publication or a more striking result for promotion. This happens in both the private sector and in academia around the world.

In 2002, an internal committee at Bell Labs found its researcher Jan Hendrik Schön, a *wunderkind* investigator into organic semiconductors, to have completely invented some of his results (published in marquee venues like *Science* and *Nature*). Schön was fired. In 2004, the University of Konstanz, in Germany, revoked his doctorate, and Schön lost his appeal. In 2006, Hwang Woo-suk, a South Korean veterinarian who made headlines by claiming he had created human embryonic stem cells by cloning, was found guilty of scientific misconduct and ethical transgressions, and the offending articles were retracted. Alongside these cases, which made headlines internationally, there are a great many other cases ranging from sloppy data collection to outright lying. It is hard to consider these as instances of pseudoscience (most scientists do not), but they bear a family resemblance.

The second phenomenon is the "replication crisis," especially prominent in the fields of experimental psychology and biomedicine. Around 2010, researchers in both fields began to notice that many of the core findings of their disciplines were difficult to reproduce in other laboratories. For many decades, scientists have regarded replication of a colleague's experiments as an important standard for reliable scientific knowledge, evidence that the results were a piece of nature rather than an artifact of a particular experimental setup. Occasionally, of course, replication proved elusive, but that was a problem for the author of the offending article; when replication failed by the dozens or the hundreds, however, it might mark a systemic crisis. The Reproducibility Project of the Open Science Collaboration in 2015 took one hundred experimental and correlational studies from three journals and was able to replicate only 36 percent. In 2012, the pharmaceutical company AmGen attempted to reproduce fifty-three "landmark" papers related to basic cancer research. It succeeded with only six. The explanations here are several, including high pressure to publish results quickly, insufficient understanding by researchers of statistics software, and conscious manipulation of statistical correlations to produce publishable papers. The extent and severity of the replication crisis remains a matter of broad debate, but the term pseudoscience does not often come up in this context.

Common among all these cases is that they came to a head *after* the publication process. The usual mechanism since the 1960s for determining whether a scientific claim is worth publishing is peer review: experts in the field evaluate the findings and either approve them or offer criticisms suggesting revision or rejection. All of these contested results cleared peer review. The volume of publication in recent decades, the limited time to replicate experiments, and the fact that most published findings are never cited—the median citation rate for an article in biomedicine is zero, meaning over half of the published literature is not utilized at all—tilt the incentives so that it might be worth it for a researcher to publish a wobbly or even fraudulent claim. The chances of getting caught are low. Peer review seems unable to catch this shoddy work, and it likely cannot catch fraud if it is executed skillfully enough. Science's communicative environment indicates that the fringe is going to continue to border the incorrect, the fraudulent, and the bad for the conceivable future.

Chapter 7
The Russian questions

It is hard to deny that humanity knows vastly more about how the universe operates than it did a millennium or a century ago, or even a decade ago. This does not mean, however, that we have nature's precise playbook, that today's mainstream scientific consensus will not be modified, perhaps even radically, in the future. But it does leave us with a puzzle when it comes to the persistence of fringe doctrines. If science brings enlightenment, and enlightenment vanquishes misguided theories, why is it so easy to find "pseudosciences" in our present scientific era?

It is easy for the same reasons that it has always been easy to find them in previous eras. We are not going to get rid of pseudoscience. Given that our theories of the natural world change over time, and that our system of knowledge production is organized around competing researchers investigating similar questions and offering their differing interpretations as superior to accepted knowledge, there will inevitably be winners and losers. Those on the losing side could simply always concede the advantages of the victorious claims and not insist on views that the consensus has now rejected, but that does not seem especially likely. The adversarial framework in which science is produced necessarily generates vestigial and controversial claims that can attract adherents.

Such vestigial and controversial cases, however, do not comprise the entirety of what the mainstream consensus dubs pseudosciences. Certain ideas about the natural world have occasionally been appropriated as tools of propaganda and ideology by political movements (or their leaders). That politicized adoption is itself a byproduct of the high prestige of the natural sciences and related to the customary expectation in the modern era that governments should support scientific research. Where there is state investment, there is state scrutiny, and sciences can become hyperpoliticized. This, too, is structural.

These are dispiriting findings. What are we to conclude about the complex landscape of fringe doctrines? It helps to break this question down into two distinct parts, which I term "the Russian questions," given that both were the titles of (not very good) nineteenth-century Russian novels: Alexander Herzen's *Who Is to Blame?*, published in 1845–1846, and Nikolai Chernyshevsky's *What Is to Be Done?*, published in 1863. (The latter title was also used by Vladimir Lenin for a political treatise in 1902.) These two questions encompass much of how we often think about the phenomenon of fringe sciences.

Who is to blame?

For as long as the word has existed, *pseudoscientist* has been a term of abuse. Nobody has ever assumed it as a self-description in earnest; it is always bestowed by opponents wanting to discredit the target. The typical answer to the blame question, proposed by those doing the labeling, is that the pseudoscientists themselves are to blame for pseudoscience. If only they would cease defending these questionable doctrines, pseudoscience would vanish.

This approach has the virtue of straightforwardness, but it comes with conceptual drawbacks that preclude it from being a solution. First, the ostensible pseudoscientists do not agree that they are defending pseudoscience; on the contrary, those involved in counterestablishment sciences often deem the mainstream consensus to be the true bastion of pseudoscience. Not only is the slur cast back from the heterodox fringe to the orthodox establishment, but it is also widely voiced among denizens of the fringe to characterize their virtual neighbors. Cosmic catastrophists consider creationists to be pseudoscientists, astrologers denigrate flat-Earthers, and so on ad nauseam.

This would be a bit easier to sort out if most members of the mainstream consensus agreed on who counted as a pseudoscientist, but that is not the case. The "consensus" turns out to be squishier than you might expect. For example, a dominant theory in cosmology at present is known as cosmic inflation, which argues that after the Big Bang the universe experienced a temporary accelerated expansion. Yet one of the initial formulators of that theory, Paul Steinhardt, in 2017 described that consensus as "pseudoscience." The advocates of inflation strongly disagree, and Steinhardt's position remains a minority one.

Consider also the case of superstring theory, a notoriously mathematically forbidding domain that seeks to unify quantum theory (the science of the very small) with general relativity (Einstein's theory of gravity, most manifest on astronomical scales) through positing tiny strings vibrating in ten or eleven dimensions. String theory's potential to realize a "theory of everything" made it enormously attractive in the mid-1980s, and by the turn of the century it was one of the chief subfields of theoretical physics. Nonetheless, the tiny scale at which the strings ostensibly operate leaves the theory largely inaccessible to empirical confirmation. Opponents have protested that its (practical) inaccessibility to experiment and its (professional) stifling of alternative approaches demand its rejection. Its advocates, on the other hand, tout its formal elegance and mathematical consistency, though they have conceded some points to their critics by shifting its moniker to "quantum gravity," thereby welcoming divergent research schools. We surely want a diversity of ideas in the sciences, but it is devilishly hard to get the balance right without tipping into either a monoculture or "anything goes." If it is hard to sharply demarcate the consensus at the cutting edge of scientific research, it is no wonder that picking out the pseudoscientists on the fringes is challenging.

Even if we could determine who the pseudoscientists were, what would we be blaming them for? Are we just condemning individuals for certain behaviors: laser-beam focus on a specific question, resistance to disconfirming evidence, tendency to long-windedness or doggedness in defending those views, a penchant for writing missives with profligate use of capital letters and exclamation points, and other ostensible hallmarks of "cranks"? Many advocates of fringe doctrines indeed display some if not all of these characteristics. The rub is that so do many scientists who work productively on behalf of the consensus using mainstream practices and theories. It is surely possible that there are personality traits shared by many of those labeled pseudoscientists, but it is also demonstrable that those characteristics are not monopolized by those on the outs. Abnormal psychology is not going to provide a neurological answer to Popper's philosophically insoluble demarcation problem.

Sometimes there is no question about whom to blame. Lysenko, Lenard, and Stark bear heavy responsibility for the travesties they promoted. But others worked in good faith to understand the natural world, albeit according to assumptions that the scientific community considered at best highly unorthodox, if not outright false. Rather than casting blame on individuals for engaging in what is, more often than not, sincere but ill-placed enthusiasm for science, we might adopt a similar strategy to that which we considered with respect to the demarcation problem. Instead of positing a global culprit, we can look for local ones, and thus avoid unwittingly creating scapegoats.

Denialism

A good example is what is sometimes called "denialism." As with "pseudoscientist," the people tagged as denialists do not embrace the label. Nonetheless, they do engage in a common set of behaviors and share personal connections that render the designation reflective of a sociological reality.

Denialism is a collection of oppositional arguments—often presented by credentialed advocates using mathematical analyses and graphs—which aims not so much to establish a proposition about the natural world, but rather to cast doubt on elements of the mainstream scientific consensus. Increasing skepticism without providing a coherent alternative is one shared characteristic of denialist practices; another is their common target: essentially all denialist claims that have emerged after World War II in the United States (and to a lesser extent in other industrialized nations) have been dedicated to vitiating government regulation of lucrative industries whose activities pose threats to the public welfare. One can find denialists prominently attacking the science behind climate change caused by the burning of fossil fuels, the destruction of the ozone layer of the upper atmosphere, the generation of acid rain, and the harms of tobacco smoking.

When we find similar behavior in a number of realms, there are a few hypotheses to explain the resemblance. It could just be coincidence. It could be that these distinct domains face analogous pressures, and these pressures have generated the independent invention of specific strategies. It could be that different critics have learned from what was happening in other domains and then adapted winning arguments. Or there could be a common source and all these seemingly different arguments are part of a coherent playbook. According to thorough research by historians of science Naomi Oreskes and Erik Conway, in the case of postwar American denialism, the correct hypothesis is the last one.

The strategy of denialism was created not by scientists but by a public relations firm. In 1954, Hill & Knowlton was hired by a client facing a crisis: the tobacco industry. Industry leaders knew that smoking cigarettes was dangerous for the health of the smoker, that the addictive properties of nicotine made it hard to quit, and that therefore their product, when used as directed, would kill its consumers. They knew this but wanted to stay in business, and so they needed a communications strategy to muddy the waters. (Executives' efforts to cover up fundamental scientific findings led to the conviction of the major U.S. tobacco companies under civil racketeering laws in 2006.) Hill & Knowlton's basic idea was not to present alternative scientific evidence but rather to cast doubt about the increasingly robust scientific consensus, to demand "more research." Who could oppose more science? As described in secret industry documents from the time, "doubt is our product."

Oreskes and Conway detailed how this process moved from industry to industry as companies wanted to avoid (or at least postpone) government regulation. The most politically visible version of this strategy was that adopted by the fossil fuels industry to combat arguments that atmospheric emissions of carbon dioxide were strongly shaping Earth's climate, raising global temperatures, and increasing ocean acidity. That carbon dioxide emissions were rising was not controversial, although whether the planet's feedback loops—absorption in the ocean and increasing cloud cover, to name two—might enhance or counter any "greenhouse effect" was, in the 1960s and 1970s, a matter of intense research. By the 1980s, there was consensus about significant warming of the planet in the coming century, threatening catastrophic consequences. Companies that profited from sales of oil and other fossil fuels wanted to defer political intervention.

The denialist strategy resembles that of counterestablishment science in the sense that it creates alternative institutions—typically, industry-funded think tanks whose corporate sponsorship is obscured to deflect public suspicion—but it sharply differs in that the denialists do not present themselves as outside establishment science. Their essential tactic is to maintain that sowing doubt, calling for more research, and muddying the public's understanding of a clear consensus in the scientific community is simply science acting normally. But there is much that is not normal here. Think tanks release findings in detailed official-seeming reports that resemble scientific literature, but the documents are commissioned without the disciplinary structures of peer review that have become customary in scientific publication. To their main audiences—politicians and the public—this distinction can be lost, which is the point. In some areas, such as regulation of coal-burning plants to decrease acid rain, or phasing out chlorofluorocarbon compounds to prevent damage to the ozone layer, the strategy delayed action for only a few years. For the harms of tobacco smoking and climate change, it has been substantially more effective. (In the case of the latter, it is still operational, although its force seems to be attenuating.)

Does this count as pseudoscience? As always, it depends on your definition, which in turn depends on how you define "science." The denialist strategists have cleverly found credentialed scientists—often politically sympathetic with antiregulatory politics—to promote views that appear, on their face, to be based on straightforward practices associated with mainstream science (more data analysis, further testing with new methods, and so on). This does not map exactly on to either hyperpoliticized science or counterestablishment science. Denialists toss allegations of pseudoscience—especially the bugbear of "Lysenkoism"—against representatives of the consensus and vice versa, and it is easy to get lost in a semantic morass. The key point is how these claims function in the public sphere: once you understand how the denialist strategy works against the public interest, the particular label matters less.

A different movement deploys many of the same tactics as the corporate-backed denialists, but originates from grass-roots community activism: anti-vaccination groups (anti-vaxx). This is a complex movement with multiple different strands. While in its contemporary variant hostility to vaccination is associated with left-wing activists concerned about child health and dates back only two decades, resistance has actually been around as long as vaccination itself has—for more than two centuries.

Eighteenth- and nineteenth-century opponents of vaccination against infectious diseases raised two chief objections: that vaccination itself carried a risk to the inoculated person, who might catch the disease from the attenuated pathogen in the serum, or suffer other damaging side effects; and that imposition of an obligation to vaccination was unwarranted government intrusion into private decisions. The first worry remains reasonable, although vaccines have become much safer in the

The Anti-Vaccination Society of America

OTHERWISE

An Association of "half-mad", "misguided" people, who write, and toil, and dream, of a time to come, when it shall be lawful to retain intact, the pure body Mother Nature gave, sends GREETING to a "suspect". "Liberty cannot be given, it must be taken."

You are Invited to Join Us

Frank D Blue, Sec'y, Terre Haute, Ind. **1902** Hon L H Piehn, President

Enclose 25c for certificate of membership.

10. Anti-vaccination organizations have a long history in the United States, dating back well before this 1902 advertisement.

intervening centuries. To counter inevitable (though rare) bad outcomes, most governments provide compensation for vaccine-related injuries. The public health benefits of widespread vaccination can be overwhelmingly demonstrated by such victories as the complete eradication of smallpox in 1980 and the almost complete elimination of polio in most regions of the globe. The consequences of an unchecked viral infection without a vaccine can be seen in the rapid and destructive spread of the novel coronavirus responsible for COVID-19 in 2020. All hopes for an end to the illness and the economic collapse induced by quarantine-based public health measures rest on an effective vaccine.

COVID-19 has evoked some conspiracy theories about its origin in China, and many proposals for dubious treatments, but the use of the term *pseudoscience* was initially quite muted. That changed as the pandemic evolved over 2020, for reasons that will seem familiar. Enormous resources and attention were suddenly focused on understanding this novel coronavirus. The particular characteristics of the virus that causes COVID-19 began to emerge, including its substantial asymptomatic period in which carriers are contagious, and its highly variable symptomology and disease outcomes. Studies flooded the Internet about the efficacy (or lack thereof) of masks, social distancing, refraining from face-touching, quarantining one's mail, various speculative treatments, and so on. As these studies were publicized and then occasionally retracted, vestigial knowledge claims proliferated, occasionally garnering numerous adherents. Others reacted with hyper-skeptical suspicion of all claims about COVID-19. Allegations of "pseudoscience" erupted. The fringing was happening between daily news cycles, an intrinsic consequence of the knowledge-production process.

To return to the anti-vaxxers: their second concern also persists, and is today associated with libertarian, survivalist, and other ideologies suspicious of government action, as well as with religious movements such as the Christian Scientists and the Dutch Reformed Church that proscribe or discourage certain medical treatments. The nonreligious groups tend to be identified with the political right and are often forgotten in public discussion of anti-vaxx. However, the lobbying of both religious and antigovernment groups induced many states within the United States to allow individuals to opt children out of obligatory vaccinations. This legal mechanism would be exploited by the anti-vaxxers of the twenty-first century.

Anti-vaxx bases its position on a 1998 article Andrew Wakefield and twelve colleagues published in the British medical journal *The Lancet*. Based on the study of a dozen children with stomach disorders, Wakefield, a gastroenterologist, asserted a correlation between the vaccine for measles, mumps, and rubella (MMR) and autism. Physicians have long been concerned by extensive evidence that autism rates are rising globally—from 1 in 2,500 in 1970 to more than 1 in 150 today—but whether this growth was due to some external factor or to increased awareness (and therefore diagnosis) continues to be a matter of heated debate. Wakefield pinpointed the MMR vaccine and thus gave parents with autistic children something to blame.

In the United States, opt-out rates from MMR rose dramatically and continued to do so even after Wakefield's *Lancet* article was retracted in 2010. The piece was also disavowed by ten of its thirteen authors, though not by Wakefield, who has turned his advocacy of an MMR-autism link into a second career. (His medical license was revoked in 2010.) Numerous studies, such as that by the National Academy of Sciences in 2001 entitled *Measles-Mumps-Rubella Vaccine and Autism*, have debunked the claim, yet anti-vaxxers continue to cite the now-discredited Wakefield publication as though it still had credibility within the scientific community. The consequences of the persistence of this superseded claim are evident: there were 1,249 measles outbreaks in the United States from January to September 2019, 89 percent of whose victims were unvaccinated or had an unknown status.

A distinctive feature of anti-vaxx as compared with other fringe movements is the prominence of women in its ranks. (This was also true of Spiritualism, although in that case the women were frequently the objects of study as well as protagonists.) The association of anti-vaxx with preventing autism in childhood, a traditional domain of women as mothers, was prefigured in the eugenics movement, which also displayed strong representation of women. This speaks to a more general pattern with fringe medicine. Those unorthodox movements claiming breakthroughs in disease prevention and treatment often appeal to women as well as men, giving those communities a different structure compared with fringe science. Responsibility for not having one's children vaccinated obviously leads to pointing fingers, but public opprobrium is unlikely to end the movement. Anti-vaxxers resemble counterestablishment movements in their ability to organize and maintain their views despite lack of access to mainstream outlets.

What is to be done?

Blaming and debunking do not prevent advocates of marginalized theories from holding fast to their views. Are there better solutions? Several proposals have been mooted over recent decades in order to counter a perceived resurgence in fringe thinking. (Whether it has in fact been on the rise is difficult to establish.)

One set of proposals has been to radically tighten our standards for scientific publication. We could, in principle, set our standards for what is publishable extremely high, so only findings very close to the scientific consensus will be endorsed. This would indeed exclude fringe ideas, but it would also stifle almost all innovation; quantum theory and continental drift would not have passed this standard. On the other hand, we could relax the standards for scientific publication to allow for some vetting—such as, for example, peer review—that would also occasionally let in unorthodox ideas. This is roughly where the bar is set now, and as a result fringe ideas do make it into the published literature, as with ESP research. Can we tinker here to improve the situation?

I think not. Suppose, for example, that we insisted on only looking at peer-reviewed research. That would rule out the think-tank reports of the denialists, but it would not exclude Wakefield (who passed peer review), and it would pose problems for today's physicists, who largely communicate by uploading unrefereed manuscripts to a preprint server. With increased specialization and mounting commitments on researchers' time, peer review seems unable to sustain the epistemic demands placed on it. It has never been very good at catching fraud, and the emergence of predatory journals (which apply no standards at all on content and survive by charging authors hefty fees), ghostwritten articles for industry, poor statistical testing (resulting in the replication crisis), and so on, further strain our scientific publication system. That said, the system seems to work well on average, even if it is not perfect. Adding obligatory statements about funding sources and conflicts of interest has also discouraged certain publications of dubious sincerity. More helpful might be relaxing the pressures for scientists to publish ever more and ever more quickly, an imposition which taxes the whole system's ability to evaluate the credibility of findings and at the same time provides incentives for fraudulent or slipshod research. Such a reform would necessarily slow the rate of scientific research (or at least scientific publication), though possibly with salutary effect.

Modifying publication directs attention to the scientists; we could instead focus on the consumers of fringe doctrines. Here we might follow the insistence of Carl Sagan and other scientists associated with the debunking CSICOP group and call for better science education. Science education is a wonderful thing, and I am entirely in favor of it. It seems unlikely, however, that improvements in scientific literacy would stamp out the fringe. Consider the flat-Earthers: every single one of them learned about the spherical shape of our planet in school, yet this has not prevented the birth of a new movement. Expansion of scientific literacy would not change the attraction of fringe doctrines for many individuals, though it might change which doctrines they found compelling—more Bigfoot, perhaps, and less alchemy.

Pseudosciences do not develop because people have insufficient scientific information. Fringe doctrines are generated through the regular process of scientific research, sloughed off from the consensus as it changes, and then gradually garner adherents. Some people join these groups for a sense of community, others because it simply makes more sense than what their science teachers tell them is the case, and others—such as the practicing scientists who stump for ESP—out of a sincere quest for the truth.

All those who have been called pseudoscientists think they are scientists. The reason they engage in these activities is not because they are anti-science, but because they are for it. Pseudoscience is the shadow of science: it is the reflection of the scientific community. The higher the status of science, the sharper the shadow and the more robust the fringe. The only way to eliminate pseudoscience is to get rid of science, and nobody wants that. What is to be done? Understanding more of the processes at work in the creation of the fringe, and its heterogeneity, helps us grapple with those few movements that can cause significant public harm. The rest we might treat as a vibrant, but mostly unthreatening, phenomenon of contemporary culture. Not all shadows hide monsters.

References

Chapter 1

Hippocrates, "The Sacred Disease," in *The Medical Works of Hippocrates*, ed. and trans. John Chadwick and W. N. Mann (Oxford: Blackwell Scientific Publications, 1950), 179.

Karl Popper, "Science: Conjectures and Refutations," in *Conjectures and Refutations: The Growth of Scientific Knowledge* (New York: Routledge, 2002 [1963]), 44, 47, 48, emphasis in original.

Larry Laudan, "The Demise of the Demarcation Problem," in *But Is It Science?: The Philosophical Question in the Creation/Evolution Controversy*, ed. Michael Ruse, updated ed. (Amherst, NY: Prometheus Books, 1988), 346.

Chapter 3

Philipp Lenard, "Foreword to 'German Physics,'" reproduced in Klaus Hentschel and Anne Hentschel, ed. and trans., *Physics and National Socialism: An Anthology of Primary Sources* (Basel: Birkhäuser Verlag, 1996), 100, 102, 103. Emphasis in original.

Stalin quoted in Nikolai Krementsov, *Stalinist Science* (Princeton, NJ: Princeton University Press, 1996), 159.

Lysenko quoted in Krementsov, *Stalinist Science*, 173. Emphasis in original.

Chapter 5

Martin Gardner, *Fads and Fallacies in the Name of Science* (New York: Dover, [1957]), 299.

James Randi, *Flim-Flam!: Psychics, ESP, Unicorns, and Other Delusions*, rev. ed. (New York: Prometheus Books, 1982), 2. Emphasis in original.

Further reading

The demarcation problem

The account presented in this book draws from Michael D. Gordin, "Myth 26: That a Clear Line of Demarcation Has Separated Science from Pseudoscience," in *Newton's Apple and Other Myths about Science*, edited by Ronald L. Numbers and Kostas Kampourakis, 219–225 (Cambridge, MA: Harvard University Press, 2015).

Every introduction to the philosophy of science will include some discussion of the demarcation problem, including: Alan Chalmers, *What Is This Thing Called Science?*, 4th ed. (Indianapolis: Hackett, 2013 [1976]); Naomi Oreskes, *Why Trust Science?* (Princeton, NJ: Princeton University Press, 2019); and Samir Okasha, *Philosophy of Science: A Very Short Introduction* (New York: Oxford University Press, 2002). Specialized philosophical accounts tend to be critical of Popper: Massimo Pigliucci, *Nonsense on Stilts: How to Tell Science from Bunk* (Chicago: University of Chicago Press, 2010); and Massimo Pigliucci and Maarten Boudry, eds., *Philosophy of Pseudoscience: Reconsidering the Demarcation Problem* (Chicago: University of Chicago Press, 2013). A sociological approach is provided in Thomas F. Gieryn, *Cultural Boundaries of Science: Credibility on the Line* (Chicago: University of Chicago Press, 1999); and Harry Collins and Trevor Pinch, *The Golem: What You Should Know about Science* (Cambridge: Cambridge University Press, 1998). Material on individual movements and doctrines can be found in the following sections.

Langmuir's "pathological science" speech is published as "Pathological Science," transcribed and edited by Robert N. Hall, *Physics Today* 42, no. 10 (1989): 36–48.

Astrology

The literature on the history of astrology is vast and generally highly specialized. For the history up to the Renaissance, good points of entry are Anthony Grafton, *Cardano's Cosmos: The Worlds and Works of a Renaissance Astrologer* (Cambridge, MA: Harvard University Press, 1999); Darin Hayton, *The Crown and the Cosmos: Astrology and the Politics of Maximilian I* (Pittsburgh: University of Pittsburgh Press, 2015); and the many mentions throughout David C. Lindberg, *The Beginnings of Western Science: The European Scientific Tradition in Philosophical, Religious, and Institutional Context, Prehistory to A.D. 1450*, 2d. ed. (Chicago: University of Chicago Press, 2007). On astrology and weather prediction in the nineteenth century, see Katharine Anderson, *Predicting the Weather: Victorians and the Science of Meteorology* (Chicago: University of Chicago Press, 2005).

Alchemy

Scholarship on the history of alchemy has grown tremendously in recent decades. Good places to start are Lawrence M. Principe, *The Secrets of Alchemy* (Chicago: University of Chicago Press, 2013); William R. Newman and Lawrence M. Principe, *Alchemy Tried in the Fire: Starkey, Boyle, and the Fate of Helmontian Chymistry* (Chicago: University of Chicago Press, 2002); and Bruce T. Moran, *Distilling Knowledge: Alchemy, Chemistry, and the Scientific Revolution* (Cambridge, MA: Harvard University Press, 2005).

Science and National Socialism

Alan D. Beyerchen, *Scientists under Hitler: Politics and the Physics Community in the Third Reich* (New Haven CT: Yale University Press, 1977); Mark Walker, *Nazi Science: Myth, Truth, and the German Atomic Bomb* (New York: Plenum Press, 1995); and Klaus Hentschel and Anne Hentschel, ed. and tr., *Physics and National Socialism: An Anthology of Primary Sources* (Basel: Birkhäuser Verlag, 1996).

Lysenkoism

Nikolai Krementsov, *Stalinist Science* (Princeton, NJ: Princeton University Press, 1997); David Joravsky, *The Lysenko Affair* (Cambridge, MA: Harvard University Press, 1970); Nils Roll-Hansen, *The Lysenko Effect: The Politics of Science* (Amherst, NY: Humanity Books, 2005); and Loren R. Graham, *Science in Russia and the Soviet Union: A Short History* (Cambridge: Cambridge University Press, 1992).

Eugenics and racial science

Daniel J. Kevles, *In the Name of Eugenics: Genetics and the Uses of Human Heredity* (New York: Knopf, 1985); Nathaniel Comfort, *The Science of Human Perfection: How Genes Became the Heart of American Medicine* (New Haven, CT: Yale University Press, 2012); and Philippa Levine, *Eugenics: A Very Short Introduction* (New York: Oxford University Press, 2017).

Phrenology

Roger Cooter, *The Cultural Meaning of Popular Science: Phrenology and the Organization of Consent in Nineteenth-Century Britain* (Cambridge: Cambridge University Press, 1984); and James Poskett, *Materials of the Mind: Phrenology, Race, and the Global History of Science, 1815–1920* (Chicago: University of Chicago Press, 2019).

Creationism

Ronald L. Numbers, *The Creationists: From Scientific Creationism to Intelligent Design* (Cambridge, MA: Harvard University Press, 2006); Robert T. Pennock and Michael Ruse, eds., *But Is It Science?: The Philosophical Question in the Creation/Evolution Controversy* (Amherst, NY: Prometheus Books, 2009); and Christopher P. Toumey, *God's Own Scientists: Creationists in a Secular World* (New Brunswick, NJ: Rutgers University Press, 1994). On the Scopes trial, see Edward J. Larson, *Summer for the Gods: The Scopes Trial and America's Continuing Debate over Science and Religion* (New York: Basic Books, 2006).

Cryptozoology

The literature on cryptozoology is divided into case studies by creature. This chapter draws primarily from Joshua Blu Buhs, *Bigfoot: The Life and Times of a Legend* (Chicago: University of Chicago Press, 2009); and Henry H. Bauer, *The Enigma of Loch Ness: Making Sense of a Mystery* (Urbana: University of Illinois Press, 1986). The latter is written by an engineer well versed in the literature of science studies who is also a believer in Nessie. Another useful history is Brian Regal and Frank J. Esposito, *The Secret History of the Jersey Devil: How Quakers, Hucksters, and Benjamin Franklin Created a Monster* (Baltimore: Johns Hopkins University Press, 2018).

Cosmic catastrophism and ancient aliens

On Velikovsky: Michael D. Gordin, *The Pseudoscience Wars: Immanuel Velikovsky and the Birth of the Modern Fringe* (Chicago: University of Chicago Press, 2012), which also contains references about Erich von Däniken.

Ufology

Greg Eghigian, "Making UFOs Make Sense: Ufology, Science, and the History of Their Mutual Misunderstanding," *Public Understanding of Science* 26 (2017): 612–626.

Flat earth

Daniel J. Clark, dir., *Behind the Curve* (2018), available at https://www.behindthecurvefilm.com; Jeffrey Burton Russell, *Inventing the Flat Earth: Columbus and Modern Historians* (New York: Praeger, 1991); Christine Garwood, *Flat Earth: The History of an Infamous Idea* (New York: St. Martin's Press, 2007); and L. Sprague de Camp, *Lost Continents: The Atlantis Theme in History, Science, and Literature* (New York: Dover, 1970 [1954]). Survey data cited from https://www.cnn.com/2019/11/16/us/flat-earth-conference-conspiracy-theories-scli-intl/index.html.

Mesmerism

Robert Darnton, *Mesmerism and the End of the Enlightenment in France* (Cambridge, MA: Harvard University Press, 1968); Jessica Riskin, *Science in an Age of Sensibility: The Sentimental Empiricists of the French Enlightenment* (Chicago: University of Chicago Press, 2002), chapter 6; and Alison Winter, *Mesmerized: Powers of Mind in Victorian Britain* (Chicago: University of Chicago Press, 1998).

Spiritualism

Janet Oppenheim, *The Other World: Spiritualism and Psychical Research in England, 1850–1914* (Cambridge: Cambridge University Press, 1985); Richard Noakes, *Physics and Psychics: The Occult and the Sciences in Modern Britain* (Cambridge: Cambridge University Press, 2019).

ESP and debunking

Seymour H. Mauskopf and Michael R. McVaugh, *The Elusive Science: Origins of Experimental Psychical Research* (Baltimore: Johns Hopkins University Press, 1980); Ian Hacking, "Telepathy: Origins of Randomization in Experimental Design," *Isis* 79, no. 3 (September 1988): 427–451; H. M. Collins and T. J. Pinch, *Frames of Meaning: The Social Construction of Extraordinary Science* (Boston: Routledge & Kegan Paul, 1982); and David Kaiser, *How the Hippies Saved Physics: Science, Counterculture, and the Quantum Revival* (New York: Norton, 2011).

Polywater, water memory, and cold fusion

On polywater: Felix Franks, *Polywater* (Cambridge, MA: MIT Press, 1981). Benveniste and cold fusion are both treated in chapter 8 of Melinda Baldwin, *Making* Nature*: The History of a Scientific Journal* (Chicago: University of Chicago Press, 2015), and cold fusion is mentioned in many of the general accounts listed earlier for the demarcation problem. Detailed accounts include Bruce V. Lewinstein, "Cold Fusion and Hot History," *Osiris* 7 (1992): 135–163; Frank Close, *Too Hot to Handle: The Story of the Race for Cold Fusion* (London: W. H. Allen, 1990); and Bart Simon,

Undead Science: Science Studies and the Afterlife of Cold Fusion (New Brunswick, NJ: Rutgers University Press, 2002).

Fraud and the replication crisis

Eugenie Samuel Reich, *Plastic Fantastic: How the Biggest Fraud in Physics Shook the Scientific World* (New York: Palgrave Macmillan, 2009); Nicolas Chevassus-au-Louis, *Fraud in the Lab: The High Stakes of Scientific Research*, tr. Nicholas Elliott (Cambridge, MA: Harvard University Press, 2019); John P. A. Ioannidis, "Why Most Published Research Findings Are False," *PLOS Medicine* (August 30, 2005), http://dx.doi.org/10.1371/journal.pmed.0020124; and Susan Dominus, "When the Revolution Came for Amy Cuddy," *New York Times Magazine* (October 22, 2017).

Denialism

The key reference is Naomi Oreskes and Erik M. Conway, *Merchants of Doubt: How a Handful of Scientists Obscured the Truth on Issues from Tobacco Smoke to Global Warming* (New York: Bloomsbury, 2010). The account of anti-vaxx draws from Michael Specter, *Denialism: How Irrational Thinking Hinders Scientific Progress, Harms the Planet, and Threatens Our Lives* (New York: Penguin, 2009), chapter 2. On politicization and scientific literacy, see Chris Mooney, *The Republican War on Science* (New York: Basic Books, 2005); and Chris Mooney and Sheril Kirshenbaum, *Unscientific America: How Scientific Illiteracy Threatens Our Future* (New York: Basic Books, 2009).

Index

A

B

C

Index

D

E

F

K

L

M

N

O

P

Q

R

S

T

U

V

W

Y

Z